Lecture Notes in Mathematics

A collection of informal reports and seminars
Edited by A. Dold, Heidelberg and B. Eckmann, Zürich

W0225792

167

Lavrentiev · Romanov Vasiliev

Computing Center of the Academy of Sciences
Novosibirsk / UdSSR

Multidimensional Inverse Problems for Differential Equations

Springer-Verlag
Berlin · Heidelberg · New York 1970

ISBN 3-540-05282-8 Springer Verlag Berlin · Heidelberg · New York
ISBN 0-387-05282-8 Springer Verlag New York · Heidelberg · Berlin

© by Springer-Verlag Berlin · Heidelberg 1970. Library of Congress Catalog Card Number 70-140559.

Offsetdruck: Julius Beltz, Weinheim/Bergstr.

CONTENTS

INTRODUCTION

An inverse problem for a differential equation is any problem involving
the determination of the coefficients or right-hand side of a differential equation from certain functionals of its solution.

Two significant advances have been made previously in the study of
inverse problems for differential equations.

The first is in inverse problems for STURM-LIOUVILLE equations ([1],
[13], [19], [23], [36]). In these problems, the coefficient in a second-
order differential operator is required to be found from the spectral
function of the operator. In [19] and [2], a number of problems involving the determination of the coefficients of a partial differential
equation are shown to be reducible to inverse problems for STURM-LIOUVILLE
equations. It is assumed there that the coefficients are functions of
a single variable.

The second is in problems of potential theory ([17], [20], [25], [29],
[33]). In the inverse problems of this type, the right-hand side of an
elliptic partial differential equation has to be determined. Highly
restrictive additional conditions are imposed on the right-hand side.
Thus, in [17] and [20], the required right-hand side in POISSON'S
equation is assumed to be a function having values 0 and 1 only,
and the set where it is 1 is a star-shaped domain. The same sort of
restrictions are also imposed on the right-hand side in other papers
dealing with the inverse problem of potential theory.

Until now, multidimensional inverse problems have been given comparative-
ly little consideration. In a multidimensional problem, the required
coefficients or right-hand sides of the differential equations are gene-
rally arbitrary functions of several variables belonging to a certain
function space.

Multidimensional problems were first investigated in the papers of Ju.
M. BERZANSKII. In [4] a uniqueness theorem was proved for the solution
to the inverse problem for SCHRÖDINGER'S equation in the class of piecewise analytic functions. In [8] and [9] functions were constructed for
some multidimensional inverse problems of quantum scattering theory
that are similar to the GEL'FAND-LEVITAN functions occurring in the

inverse problem for the STURM-LIOUVILLE equation.

This monograph investigates a number of multidimensional inverse problems whose formulations differ from those of the papers mentioned above. A portion of the results have been published as short notes ([21], [22], [30], [31]).

A characteristic aspect of multidimensional inverse problems is their property of not being well-posed in the sense of HADAMARD. Thus, it is advantageous to make use of the general notions and approaches to ill-posed problems developed in [16], [20], [34], and [35]. Central to the theoretical study of a problem that is not well-posed in the sense of HADAMARD is the proof of a uniqueness theorem. The monograph consists primarily of proofs of uniqueness theorems for the formulations in question.

The inverse problems for which uniqueness theorems will be proved are linear and their solution is reduced to the solution of first-order linear operator equations. Thus algorithms may be constructed to solve them numerically by application of the general methods developed in [20] and [35] for linear equations.

The methods used to prove uniqueness lead to special algorithms for the problems. They also make possible the derivation of estimates characterizing the stability of the various formulations on certain specific compact sets (for example, the set of functions whose gradients are uniformly bounded).
We shall confine ourselves here to the study of inverse problems for second-order differential equations although some of the methods carry over to higher-order equations [18].

The inverse problems considered for hyperbolic equations are reducible to problems of integral geometry and so Chapter 1 is devoted to some aspects of it. Chapter 2 establishes uniqueness theorems for the solution to the inverse problem for the telegraph equation with the help of the results of Chapter 1 while Chapter 3 does the same thing for the wave equation.
Chapters 4 and 5 deal with inverse problems for the heat equation and for elliptic equations. They are reduced to the solution of certain integral equations of first kind.
In Chapters 3 and 4, some applied problems are discussed that lead to corresponding versions of the inverse problem.

CHAPTER 1

Some Problems of Integral Geometry

In accordance with the terminology used in [12], an integral-geometric
problem is any problem involving the determination of a function defined
in a domain through its integrals along a family of curves in the domain.
One of the earliest and most familiar versions of such problems is the
determination of a continuous function of n-variables through its mean
values over spheres of arbitrary radius with centers lying on an (n-1)-
dimensional hyperplane. The uniqueness of a solution to the problem is
proved in R. COURANT'S book [6]. At present, the determination of a
function from its integrals over hyperplanes has been the problem dealt
with most fully in monographs [12]. The authors of the book, I.M. GEL'FAND,
M.I. GRAEV and N.Ja. VILENKIN were led to its consideration while working
on problems in representation theory. The appearance of this book fur-
thered to a considerable extent the development and systematic elaboration
of integral-geometric problems. Some differently formulated problems of
integral geometry are contained in F. JOHN'S book [18].

As already indicated in the introduction, we were led to problems in
integral geometry while considering linearized inverse
problems for the simplest equations of mathematical physics. Some of
them lead to the familiar problem of determining a function from its
mean values over spheres and others to new problems which it is the
purpose of this chapter to consider. The problem of finding a function
from its integrals is not well-posed in the sense of HADAMARD. We shall
therefore proceed from A.N. TIHONOV'S notion and we shall prescribe the
function space to which the solution of a problem is to belong. With
applications of integral geometry to the study of linearized problems in
mind, the most natural one for our purposes is the space C of continuous functions. Throughout the following we shall assume the solutions
of corresponding problems in integral geometry to belong to the space
C or to some subset of it. We shall be interested in uniqueness theorems
for these problems, the determination of inversion formulas for them
and the set of functions for which these formulas are valid. The latter
is related to the question of existence of a solution.

Section 1 considers the problem of reconstructing a two-dimensional
function from its integrals over a family of ellipses with one focus

fixed and the other running over the points of a line. At the end of
the section it is shown how these results may be rephrased to encompass
the problem of reconstructing a function of n variables from its integrals over ellipsoids of revolution. In Section 2 the results of Section
1 are generalized to curves of a more general nature a special case
being that of ellipses. The formulation of the integral-geometric problem
is also generalized by the introduction of a weight factor in the integrals along the curves. Section 3 considers the determination of a function inside a circle of unit radius from its integrals over a family of
curves invariant to rotation about the center of the circle. In Section 4
an inversion formula is obtained for the problem of determining a function
from its mean values over circles. Although, as we have indicated
already, uniqueness has been proved before and a basic algorithm exists
for constructing the required function, the inversion formula has been
lacking for this problem.

1. Problem of Determining a Function from Integrals over Ellipsoids of Revolution

Consider the following problem in (n+1)-dimensional space. The integrals of a function $u(x,s) = u(x_1, x_2, \ldots, x_n, s)$ are prescribed on a
family of ellipsoids with one focus fixed at the origin and the other
running over all points of the hyperplane s = 0. Only ellipsoids obtained by revolving an ellipse around the line joining the two foci are
to be considered. It is required to determine $u(x,s)$ from the given
integrals.

Denote the coordinates of the second focus by $(x^0, 0) = (x_1^0, x_2^0, \ldots, x_n^0, 0)$
and by $S_{x^0, t}$ the ellipsoid of revolution defined by

(1) $$r(x,s,0,0) + r(x,s,x^0,0) = t$$

where $r(x,s,0,0)$ and $r(x,s,x^0,0)$ are the distances between (x,s)
and the foci $(0,0)$ and $(x^0,0)$, respectively.

Thus, let the function

(2) $$v(x^0, t) = \int_{S_{x^0, t}} u(x,s) d\omega$$

be known. Here ω is the solid angle in (x,s)-space with vertex at
the origin. In accordance with the above discussion, we shall assume

that u(x,s) belongs to the space of C-functions. From (2) it is
apparently only meaningful to pose the question of uniqueness of a
solution to this problem in the class of even functions of s and this
we shall do in the following. In addition, there is obviously no loss
of generality in assuming that u(0,0) = 0. To simplify the subsequent
computations, we shall take n = 1 so that the role of ellipsoids will
be played by ellipses and the solid angle ω will be the polar angle φ.
At the conclusion of the section, we show how the corresponding results
carry over to the case of arbitrary n.

For n = 1 , the following uniqueness theorem holds.

Theorem 1: *If equation (2) has a solution belonging to C satisfying
           a Hölder condition in a neighborhood of the origin, then it
           is unique.*

By solution here, we mean an even function of s vanishing at the origin.

The idea of the proof of the theorem is to find all moments of u(x,s).
To this end, it is convenient to go over to polar coordinates (r, φ)
related to the cartesian coordinates (x,s) by the formulas

$$(3) \qquad x = r \ \cos \varphi \ , \ s = r \ \sin \varphi \ .$$

The equation of an ellipse in polar coordinates is

$$(4) \qquad r = p(1-\varepsilon \cos \varphi)^{-1}$$

where p and ε are parameters characterizing the polar distance and
eccentricity of the ellipse; they are expressible in terms of x^0
and t by

$$(4a) \qquad \varepsilon = \frac{x^0}{t} \ , \ p = \frac{t}{2}(1 - \varepsilon^2) \ .$$

Formula (2) then becomes

$$(5) \qquad \int_0^{2\pi} u(r \cos \varphi, r \sin \varphi)d\varphi = v(p,\varepsilon) \ ,$$

with r given by (4). We apply to both sides of (5) the operator L
defined by

$$(6) \qquad Lv = p \ \frac{\partial}{\partial \varepsilon} \int_0^p v(z,\varepsilon) \ \frac{dz}{z} \ .$$

When equation (5) has a solution in C , both its legitimacy and the result of applying it are substantiated by the following sequence of equations :

$$Lv = p \frac{\partial}{\partial \epsilon} \int_0^p \frac{dz}{z} \int_0^{2\pi} u(r_z \cos\varphi, \; r_z \sin\varphi) d\varphi$$

$$= p \frac{\partial}{\partial \epsilon} \int_0^{2\pi} d\varphi \int_0^p u(r_z \cos\varphi, \; r_z \sin\varphi) \frac{dz}{z}$$

(6a)

$$= p \int_0^{2\pi} d\varphi \frac{\partial}{\partial \epsilon} \int_0^{r_p} u(r \cos\varphi, \; r \sin\varphi) \frac{dr}{r}$$

$$= \int_0^{2\pi} u(r_p \cos\varphi, \; r_p \sin\varphi) \; r_p \cos\varphi \, d\varphi = \int_{S_{p,\epsilon}} u(x,s) x \, d\varphi \quad .$$

The subscripts p and z on r indicate which of these two parameters is to be substituted in formula (4) when calculating   r. $S_{p,\epsilon}$ denotes an ellipse with parameters p and ϵ .

Applying the operator L repeatedly to the resultant equation

(7)
$$\int_{S_{p,\epsilon}} u(x,s) x \, d\varphi = Lv$$

and denoting by L^k the k-fold iteration of L, we obtain in a similar way

(8)
$$\int_{S_{p,\epsilon}} u(x,s) x^k d\varphi = L^k v$$
$$(k = 1,2,3,\dots)$$

If we define $L^0 v = v(p,\epsilon)$, formula (8) is also valid for k = 0 . Thus we have constructed in unique fashion a system of moments on each ellipse. Since u(x,s) is an even function of s, it is uniquely determined by these moments. In other words, if a solution to equation (4) exists, it is unique.

We next consider how relations (8) may be used to express the function u(x,s) explicitly in terms of v(p,ε). At the same time, we shall study the properties needed by $v(p,\varepsilon)$ to assure the existence of a solution to equation (5). At this juncture, we shall slightly contract the class of functions for which we have proved the uniqueness of the reconstructed function from its integrals over ellipses. Namely, we shall consider functions $u(r,\varphi)$ satisfying the following conditions :

1°. each $u(r,\varphi)$ is a continuous function of its arguments in the disc $r \leq r_{o}$, it is even in φ and $u(0,\varphi) = 0$. Here r_{o} is an arbitrary positive number.

2°. In a neighborhood of the polar origin, $u(r,\varphi)$ satisfies a HÖLDER condition,

$$(9) \qquad |u(r,\varphi)| \leq Ar^{\mu}, \ (\mu > 0)$$

where A and μ are constants.

3°. Each $u(r,\varphi)$ satisfies the inequality

$$(10) \qquad \sum_{k=0}^{\infty} \max_{r} |u_{k}(r)| < \infty \quad ,$$

wherein

$$(11) \qquad \begin{cases} u_{k}(r) = \dfrac{1}{\pi} \displaystyle\int_{0}^{2\pi} u(r,\varphi) \cos k\varphi\, d\varphi, & (k = 1,2,3,\ldots) \\[3mm] u_{0}(r) = \dfrac{1}{2\pi} \displaystyle\int_{0}^{2\pi} u(r,\varphi)\, d\varphi . \end{cases}$$

The functions for which conditions $1^{o}-3^{o}$ hold will be designated as class U.

We shall also make a slight change in the statement of the problem. Consider a circle of radius r_{o} in the (x,s)-plane and all ellipses of eccentricity $0 \leq \varepsilon < 1$ falling inside the circle. It is required to determine a function $u(r,\varphi) \in U$ from its integrals over this family of ellipses.

Let $v(p,\varepsilon)$ be a function for which a solution to (5) exists. Let the parameter ε tend to zero in (8). Each ellipse becomes a circle of radius p and the following relation results in the limit :

$$(12) \qquad \int_{0}^{2\pi} u(p,\varphi) \cos^{k}\varphi\, d\varphi = p^{-k}[L^{k}v]_{\varepsilon=0} . \qquad (k = 0,1,2,\ldots)$$

For each fixed value of r, these relations alone can now be used to
construct a Fourier series for $u(r,\varphi)$ which is convergent for any
$u \in U$ by virtue of condition 3°. Thus knowing the integrals along
ellipses having eccentricity ranging in the interval $0 \leq \epsilon \leq \delta$ where δ is an
arbitrarily small positive number, we can determine $u(r,\varphi)$ in the
disc $r \leq r_0$ and hence we can find its integrals along all ellipses
lying in this disc. This means that if we prescribe the function
$v(p,\epsilon)$ in an arbitrarily small strip $0 \leq \epsilon \leq \delta$, we determine it
completely for $\delta \leq \epsilon < 1$. This implies in turn that to any arbitrary
continuous function there exists no solution to the stated problem.
This result relates to the fact that the given problem is not well-posed
in the sense of HADAMARD. Actually no closed linear manifold of functions
belonging to the function spaces C^k, H, L_p or $W_p^{(\ell)}$ possesses the
property obtained above for $v(p,\epsilon)$. Below we shall indicate what properties the functions $u(r,\varphi)$ must possess that are necessary and
sufficient for the existence of a $u(r,\varphi)$ satisfying equation (5).

Consider a family of linear operators M_k defined by the relations

$$(13) \qquad M_k v = \frac{1}{\pi} \sum_{j=0}^{[\frac{k}{2}]} (-1)^j (\sum_{\ell=0}^{[\frac{k}{2}]-j} \binom{k}{2(j+\ell)} \binom{j+\ell}{\ell}) \, p^{2j-k} L^{k-2j} v,$$

$$(k=1,2,\ldots)$$

and

$$(13a) \qquad M_0 v = \frac{1}{2\pi} L^0 v - \frac{1}{2\pi} v(p,\epsilon) .$$

Using the system of relations (8), we find that the application of M_k
to (5) results in the formula

$$(14) \qquad \frac{1}{2\pi} \int_{S_{p,\epsilon}} u(r,\varphi) \{ [t+\sqrt{t^2-1}]^k + [t-\sqrt{t^2-1}]^k \} \Big|_{t=\frac{\cos\varphi}{1-\epsilon\cos\varphi}} d\varphi = M_k v .$$

It should be noted that for $|t| < 1$ the expressions in brackets are
complex. Nevertheless, the entire expression in braces is real. For
$\epsilon = 0$, equation (14) can be written as

$$(14') \qquad \frac{1}{\pi} \int_0^{2\pi} u(r,\varphi) \cos k\varphi \, d\varphi = [M_k v]_{\epsilon=0} .$$

Hence, if a solution to (5) exists which belongs to U, it is expressible
in terms of $v(p,\epsilon)$ through the formula

$$(15) \qquad u(p,\varphi) = \sum_{k=0}^{\infty} [M_k v]_{\epsilon=0} \cos k\varphi .$$

The convergence of this series for any $u \in U$ follows from $(14')$ and condition 3° on the function $u(r, \varphi)$. We now examine what properties are possessed by V, the image of the set U under the correspondence definable by (5).

Theorem 2: *The image V of U under (5) has the following properties:*

1°. The functions $M_k v$ ($k = 0,1,2,\ldots$) corresponding to a $v(p,\varepsilon) \in V$ by (13) exist and are continuous, and $[M_k v]_{p=0} = 0$.

2°. For any $v(p,\varepsilon)$ the series

$$(16) \qquad \sum_{k=0}^{\infty} \max_{p} |M_k v|_{\varepsilon=0} = \alpha_v$$

is convergent.

3°. The function $u(r, \varphi)$ constructed for $v(p,\varepsilon)$ from formula (15) satisfies the HÖLDER condition,

$$(17) \qquad |u(r,\varphi)| \leq A r^{\mu}, \; (\mu > 0) .$$

4°. Each $v(p,\varepsilon)$ satisfies the inequality

$$(18) \qquad |v(p,\varepsilon)| \leq 2\pi \, \alpha_v .$$

Properties 1°-3° follow in a trivial way from the corresponding properties of the functions $u(r,\varphi)$ and equations (14), $(14')$ and (15). Only inequality (18) remains to be proved. To this end, note that the function $u(r, \varphi)$ given by (15) has to satisfy equation (5), i.e., the following identity must hold:

$$(18a) \qquad \int_0^{2\pi} \sum_{k=0}^{\infty} [M_k v]_{\substack{\varepsilon=0 \\ p \to r}} \cos k\varphi \, d\varphi = v(p,\varepsilon) .$$

$[M_k v]_{\substack{\varepsilon=0 \\ p \to r}}$ should be interpreted to mean that $M_k v$ is first to be evaluated at $\varepsilon = 0$ and then p replaced by r as given by (4). From this identity we deduce that

$$(18b) \qquad |v(p,\varepsilon)| \leq 2\pi \sum_{k=0}^{\infty} \max_{p} |M_k v|_{\varepsilon=0} = 2\pi \, \alpha_v$$

and the theorem is proved.

Theorem 3: *A necessary and sufficient condition for equation (5) to*
have a solution belonging to U is that v(p,ε) belong to V.

The necessity follows from Theorem 2. Let us show that the belonging of
v(p,ε) to V is sufficient for the existence of a solution. Indeed,
consider for such a v(p,ε) the series

$$(19) \qquad \sum_{k=0}^{\infty} [M_k v]_{\varepsilon=0} \cos k\varphi = u(p,\varphi) \quad .$$

By virtue of property 2°, it is uniformly convergent and defines a
continuous function u(p,φ) of p and φ. Moreover, the FOURIER
coefficients for u(p,φ) coincide with $[M_k v]_{\varepsilon=0}$ (k = 0,1,2,...).
By properties 1°-3° of v(p,ε), the function u∈U (clearly, u is
even in φ and u(0,φ) = 0). It remains to show that it satisfies
equation (5). On the basis of u(p,φ) construct the function

$$(20) \qquad \tilde{v}(p,\varepsilon) = \int_{S_{p,\varepsilon}} u(r,\varphi) d\varphi \quad .$$

Evidently, $\tilde{v} \in V$. Applying M_k to equation (20), we obtain

$$(20a) \qquad [M_k \tilde{v}]_{\varepsilon=0} = [M_k v]_{\varepsilon=0} \quad .$$

Or by the linearity of M_k ,

$$(20b) \qquad [M_k w]_{\varepsilon=0} = 0, \qquad (k = 0,1,2,...)$$
$$w = v - \tilde{v} \qquad .$$

The function w is in V. Using inequality (18), we find from this that

$$(20c) \qquad w(p,\varepsilon) = 0$$

or equivalently, $\tilde{v}(p,\varepsilon) = v(p,\varepsilon)$. This means that each function v(p,ε)
produces a solution to equation (5) through formula (19). By Theorem 1
it is unique.

We return now to the case of arbitrary n and we outline how the results
obtained for ellipses carry over to the case of ellipsoids of revolution.
Thus, let $S_{x^0,t}$ be a family of ellipsoids of revolution one of whose
foci is at the origin and the other at any point of hyperplane s = 0.
Knowing the integrals of a function u(x,s) over these ellipsoids, one

is required to determine it. In other words, it is necessary to solve equation (2). The uniqueness of a solution to this equation is demonstrated just as in the case of ellipses by finding the moments of the function. To this end, we perform an orthogonal coordinate transformation in (x,s)-space with matrix Q amounting to a rotation about the origin in the hyperplane $s = 0$. Suppose that under this transformation the point (x,s) goes into (y,s) and the point $(x^o,0)$ goes into $(y^o,0)$, where $(y,s) = (y_1,y_2,...y_n,s)$ and $(y^o,0) = (y_1^o,0,...,0,0)$. The matrix of the transformation depends on several parameters which may be taken to be, in particular, the direction cosines $q_1,q_2,....q_n$ of the radius vector to the point $(x^o,0)$. Introduce spherical coordinates for y and s by the formulas

$$(21) \qquad y_1 = r\xi_1 \ , \ s = r\xi_{n+1} \ ,$$

$$(i = 1,2,...,n)$$

where $\xi_i (i = 1,2,...,n+1)$ are the direction cosines of the radius vector r in (y,s)-space. The equation of an ellipsoid of revolution can then be written in the form

$$(22) \qquad r = p(1 - \varepsilon\xi_1)^{-1}$$

where p and ε are given by

$$(22a) \qquad \varepsilon = \frac{y_1^o}{t} \ , \ p = \frac{t}{2}(1 - \varepsilon^2) \ .$$

Equation (2) can now be rewritten as

$$(23) \qquad \int_{S_{q,p,\varepsilon}} u(rQ\cdot\xi)d\omega = v(q,p,\varepsilon) \ ,$$

where $\xi = (\xi_1,\xi_2,...,\xi_{n+1})$, $q = (q_1,q_2,...,q_n)$ and $S_{q,p,\varepsilon}$ is the surface of the ellipsoid of revolution with parameters q,p and ε. Just as in the preceding, by applying the operator L^k to equation (23) keeping q fixed, where L is the operator defined by (6), we obtain

$$(24) \qquad \int_{S_{p,q,\varepsilon}} u(rQ\cdot\xi)y_1^k d\omega = L^k v$$

$$(k = 0,1,2,...) \ .$$

Note that by virtue of the orthogonality of the transformation,

$$(24a) \qquad y_1 y_1^o = x_1 x_1^o + x_2 x_2^o + ... + x_n x_n^o \ .$$

Since $(x^0,0)$ is an arbitrary point of the hyperplane $s = 0$, by taking the invariance of $d\omega$ into consideration, one can use equation (24) for $\epsilon = 0$ to derive the following moments for $u(x,s)$:

$$(24b) \qquad \int_{S_{0,t}} u(x,s) \; x_1^{\lambda_1} \cdot x_2^{\lambda_2} \ldots x_n^{\lambda_n} \; d\omega,$$

$$(\lambda_1 = 0,1,2,\ldots), \; (1 = 1,2,\ldots,n)$$

where $S_{0,t}$ is a sphere of radius t with center at the origin. These moments determine our even function of s uniquely. Using the resultant system of moments, one can construct an inversion formula in a similar way to the case $n = 1$.

2. Generalization to Analytic Curves

The method of determining a function from its integrals discussed in the preceding section can easily be generalized to surfaces of revolution more general than ellipsoids. However, we shall confine ourselves to the case $n = 1$, in other words, plane curves. Moreover, we shall only consider curves symmetric about the x-axis ($s = 0$). However, as in the preceding section, the entire discussion easily carries over to $(n+1)$-dimensional surfaces resulting by way of rotating a plane curve around an axis of symmetry.

Consider a two-parameter family of curves $S_{p,\epsilon}$ given in polar coordinates by

$$(1) \qquad r = pf(\epsilon,\epsilon \cos\varphi),$$

where $f(\epsilon,\eta)$ is an analytic function of ϵ and η in an arbitrary small neighborhood of the origin such that $f(0,0) \neq 0$ and $\frac{\partial}{\partial\eta} f(0,0) \neq 0$. Let the integrals

$$(2) \qquad \int_{S_{p,\epsilon}} \phi(\epsilon,\epsilon \cos\varphi) \; u(r,\varphi) d\varphi = v(p,\epsilon),$$

be given along these curves, wherein $\phi(\epsilon,\eta)$ is a known analytic function of ϵ and η in a neighborhood of the origin such that $\phi(0,0) \neq 0$. From formula (2), it is apparently only meaningful to consider determining uniquely from $v(p,\epsilon)$ a function $u(r,\varphi)$ even in φ. Therefore throughout the following when we speak of a solution to (2) we shall mean an even function of φ.

The following theorem holds.

Theorem 4: *If equation (2) has a solution that is bounded everywhere, belongs to C and satisfies a HÖLDER condition at the origin, then it is unique.*

The method of proving this theorem is similar to that of Theorem 1 and so we shall find the moments of $u(r, \varphi)$ on each circle $S_{p,o}$.

Expanding $\phi(\epsilon, \eta)$ in a series with respect to η and substituting it in formula (2), we obtain

$$(3) \qquad v(p, \epsilon) = \sum_{k=0}^{\infty} \epsilon^k a_k(\epsilon) \, v_k(p, \epsilon),$$

where

$$(4) \qquad v_k(p, \epsilon) = \int_{S_{p,\epsilon}} u(r, \varphi) \, \cos^k \varphi \, d\varphi,$$

$$(5) \qquad a_k(\epsilon) = \frac{1}{k!} \left. \frac{\partial^k \phi}{\partial \eta^k} \right|_{\eta=0}$$

If we are able to determine the functions $v_k(p, 0)$ uniquely in terms of $v(p, \epsilon)$, this will prove that the reconstruction of $u(r, \varphi)$ is unique. From (3), we find that

$$(5a) \qquad v_o(p, 0) = \frac{v(p, 0)}{a_o(0)} \quad .$$

Introduce the operator L defined by

$$(6) \qquad Lv = \frac{\partial}{\partial \epsilon} \int_0^p v(z, \epsilon) \, \frac{dz}{z}$$

and apply it to (4). Its validity is justified by the HÖLDER condition for the function $u(r, \varphi)$. Indeed,

$$(6a) \qquad Lv_k = \frac{\partial}{\partial \epsilon} \int_0^p \frac{dz}{z} \int_{S_{z,\epsilon}} u(r, \varphi) \, \cos^k \varphi d\varphi = \frac{\partial}{\partial \epsilon} \int_{S_{p,\epsilon}} \cos^k \varphi d\varphi \int_0^{pf(\epsilon, \epsilon \cos \varphi)} u(r, \varphi) \, \frac{dr}{r}$$

$$= \int_{S_{p,\epsilon}} u(r, \varphi) \, \cos^k \varphi \left[\frac{\frac{\partial}{\partial \epsilon} f(\epsilon, \epsilon \cos \varphi)}{f(\epsilon, \epsilon \cos \varphi)} \right] d\varphi \quad .$$

Expanding the expression in brackets in the integrand in powers of $\varepsilon \cos \varphi$, we obtain

$$(7) \qquad Lv_k = b_o(\varepsilon) \, v_k(p,\varepsilon) + \sum_{n=1}^{\infty} \left[c_n(\varepsilon) + \varepsilon b_n(\varepsilon) \right] \varepsilon^{n-1} v_{k+n}(p,\varepsilon),$$

wherein $b_n(\varepsilon)$ and $c_n(\varepsilon)$ are expressible in terms of the partial derivatives of $f(\varepsilon,n)$. By virtue of the conditions imposed on the latter function, $c_1(0) \neq 0$. The last relation allows us to now derive an equation for $v_1(p,0)$. Apply L to (3) and use formula (7) for Lv_k. This results in

$$(8) \qquad \begin{aligned} Lv = {} & d_o(\varepsilon) \, v_o(p,\varepsilon) + \sum_{k=1}^{\infty} \varepsilon^{k-1} d_k(\varepsilon) \, v_k(p,\varepsilon) \\ & + \ell_o(\varepsilon) \int_0^p v_o(z,\varepsilon) \, \frac{dz}{z} + \sum_{k=1}^{\infty} \varepsilon^{k-1} \ell_k(\varepsilon) \int_0^p v_k(p,\varepsilon) \, \frac{dz}{z} \,, \end{aligned}$$

in which $d_1(0) = a_o(0)c_1(0) \neq 0$. Letting $\varepsilon \to 0$ in equation (8) and dividing by $d_1(0)$, we arrive at the following VOLTERRA equation for $v_1(p,0)$:

$$(9) \qquad v_1(p,0) + \lambda \int_0^p v_1(\xi,0) \, \frac{d\xi}{\xi} = f_1(p)$$

λ being some numerical parameter. The equation is meaningful in that $v_1(p,0)$ satisfies a HÖLDER condition at $p=0$. Equation (9) does not have a unique solution in general since it has an eigenfunction of the form $Cp^{-\lambda}$ if $\lambda < 0$, C being an arbitrary constant. The equation has no other eigenfunctions. Note however that the boundedness of $u(r,\varphi)$ together with (4) implies the boundedness of all $v_k(p,\varepsilon)$ and so the solution we seek for (9) has to be bounded. But since for $\lambda < 0$, $Cp^{-\lambda}$ tends to infinity as $p \to \infty$ if $C \neq 0$, we can assert that any bounded solution to (9) is unique.

If we apply L to equation (3) k times in succession, we let ε tend to zero and we divide by the non-zero coefficient $a_o(0)c_1^k(0)$, we obtain for the function $w_k(z) \equiv v_k(\exp z, 0)$ a VOLTERRA integral equation of the form

(9a)
$$w_k(z) + \int_{-\infty}^{z} \left[\lambda_o(z-\xi)^{k-1} + \lambda_1(z-\xi)^{k-2} + \ldots + \lambda_{k-2}(z-\xi) + \lambda_{k-1} \right] w_k(\xi) = f_k(z),$$

where $\lambda_o, \lambda_1, \ldots, \lambda_{k-2}$ and λ_{k-1} are certain numerical coefficients and $f_k(z)$ is a continuous function expressible in terms of the functions $\left[L^j v \right]_{\varepsilon=0}$ $(j = 0,1,2,\ldots,k)$. It is easy to show by reducing

the VOLTERRA equation to a differential equation with constant coefficients that there are at most k linearly independent eigenfunctions
each of which tends to infinity as z → ∞. Hence any bounded solution
of the integral equation is unique. Thus all of the functions $v_k(p,0)$
(k = 0,1,2,...) can be found in a unique way. This implies the uniqueness of a solution to equation (2). An inversion formula can be derived
in the class of functions having convergent FOURIER series by proceeding
in the same way as in Sec.1.

The following example shows that nothing significant could be gained
if the hypothesis of the theorem concerning boundedness of $u(r,\varphi)$
could be relaxed. Even for our earlier considered case of ellipses, if
a solution is unbounded, it is possible to choose a weight function in
certain situations in such a way that the required function will not be
uniquely determined by its integrals.

Indeed, consider the function

$$(10) \qquad u(r,\varphi) = r^\gamma \sum_{k=1}^{N} A_k \cos k\varphi,$$

where γ is a positive number, N is a positive integer and the A_k
are arbitrary numerical coefficients. Choose as weight function
$\Phi(\varepsilon,\eta) \equiv (1-\eta)^\gamma$.

Then if $S_{p,\varepsilon}$ is the ellipse with parameters p and ε given by
equation (4) of Sec.1, we have

$$(10a) \qquad \int_{S_{p,\varepsilon}} (1-\varepsilon \cos\varphi)^\gamma u(r,\varphi)d\varphi = 0$$

for any p and $0 \le \varepsilon < 1$. But this means that equation (2) has corresponding to the function $v(p,\varepsilon) \equiv 0$ not only the trivial solution but
also the nontrivial solution given by (10).

3. Problem of Determining a Function inside a Circle from Integrals on a Family of Curves Invariant to Rotation around the Center of the Circle

In this section we shall consider the following problem: A continuous
function is defined inside the unit circle and its integrals along a
two-parameter family of curves are given. As before two questions are
of interest. The first is whether the function determined through these

integrals is unique and second if the function is unique how to construct it.

Introduce polar coordinates (r, φ) with the pole situated at the center of the circle.

Consider a two-parameter family of curves having the following properties:

1°. The family is invariant to rotation about the center of the circle.

2°. Each curve begins and ends on the unit circle.

3°. Each curve consists of two branches whose equations are expressible in the form

$$(1) \qquad \varphi_j = \alpha - (-1)^j \psi_j\ (r,\rho)\ \sqrt{r-\rho}\ ,$$

$$j = 1,2 \qquad (0 < \rho \leq r \leq 1) \quad .$$

Here ρ is the distance from the center of the circle to the point of the curve closest to the center of the circle (we shall henceforth refer to this point as the certex of the curve) and α is the polar angle of the point. Thus each curve of the given family is characterized by the coordinates of its vertex. Concerning the functions $\psi_j(r,\rho)$ we shall assume them to be continuous in r and ρ for $0 < \rho \leq r \leq 1$

and $\psi_1(\rho,\rho) = \psi_2(\rho,\rho) \neq 0$. The functions $\sqrt{r-\beta}\ \dfrac{\partial \psi_j}{\partial r}$ and $\sqrt{(r-\beta)^3}\ \dfrac{\partial^2 \psi_j}{\partial r \partial \rho}$

will also be assumed continuous in this same region.

Observe that a twice continuously differentiable function of arclength whose graph intersects each circle r = const. in at most two points and whose vertex is at a distance ρ from the center of the circle can be represented by an equation such as (1) only if the center of the osculating circle at the vertex does not coincide with the center of the unit circle. In the latter case, the factor $(r-\rho)^{\frac{1}{2}}$ multiplying $\psi_j(r,\rho)$ may be replaced by $(r-\rho)^{\mu}$ $(0 < \mu < \frac{1}{2})$ under certain additional smoothness conditions on the curve. However if one considers piecewise twice differentiable functions for which the vertices are corners of the curves so that there exist two-sided tangents, then μ may also assume values greater than $\frac{1}{2}$. We shall not touch upon these cases although from our considerations below it will be clear how to proceed with them.

The continuity of the functions $\psi_j(r,\rho)$ and $\sqrt{(r-\rho)^3}\ \dfrac{\partial^2 \psi_j}{\partial r \partial \rho}$ with respect to ρ are additional restrictions on the density with which the curves of the family cover the unit circle.

Thus suppose that the integrals of $u(r,\varphi)$ with respect to arclength are known on the family of curves satisfying conditions 1°-3° :

$$(2) \qquad \sum_{j=1}^{2} \int_{\rho}^{1} u(r,\varphi_j)\ \sqrt{1+\left(r\ \frac{\partial \varphi_j}{\partial r}\right)^2}\ dr = v(\rho,\alpha)\ .$$

The following theorem holds :

Theorem 5: *If $u(r,\varphi)$ is continuous in the unit disc, then it can be uniquely determined from the function $v(\rho,\alpha)$.*

The theorem is proved by finding the FOURIER coefficients of $u(r,\varphi)$. Introduce the notation

$$(3) \qquad \begin{cases} u_k(r) = \dfrac{1}{2\pi} \displaystyle\int_0^{2\pi} u(r,\varphi)\ \exp(ik\varphi)\ d\varphi \\[2mm] v_k(r) = \dfrac{1}{2\pi} \displaystyle\int_0^{2\pi} v(r,\varphi)\ \exp(ik\varphi)\ d\varphi\ , \quad (k=0,\pm1,\pm2,\ldots). \end{cases}$$

Multiplying the left and right-hand sides of (2) by $(2\pi)^{-1} \exp(ik\alpha)$ and integrating with respect to α from 0 to 2π, we obtain

$$v_k(\rho) = \sum_{j=1}^{2} \frac{1}{2\pi} \int_0^{2\pi} \exp(ik\alpha)\, d\alpha \int_{\rho}^{1} u(r,\varphi_j)\sqrt{1+\left(r\ \frac{\partial \varphi_j}{\partial r}\right)^2}\ dr$$

$$(3a) \qquad = \sum_{j=1}^{2} \int_{\rho}^{1} \sqrt{1+\left(r\ \frac{\partial \varphi_j}{\partial r}\right)^2}\ \left\{ \frac{1}{2\pi} \int_0^{2\pi} u(r,\varphi_j)\ \exp(ik\alpha)\ d\alpha \right\} dr$$

$$= \sum_{j=1}^{2} \int_{\rho}^{1} \sqrt{1+\left(r\ \frac{\partial \varphi_j}{\partial r}\right)^2}\ \left\{ \frac{1}{2\pi} \int_0^{2\pi} u(r,\varphi_j)\ \exp\left[ik\varphi_j+ik(-1)^j\psi_j(r,\rho)\cdot\sqrt{r-\rho}\right] d\varphi_j \right\} dr$$

$$= \sum_{j=1}^{2} \int_{\rho}^{1} u_k(r)\sqrt{1+\left(r\ \frac{\partial \varphi_j}{\partial r}\right)^2}\ \exp\left[ik(-1)^j\psi_j(r,\rho)\sqrt{r-\rho}\right]\ dr\ \ .$$

Thus the function $u_k(r)$ satisfies the equation

$$(4) \qquad \int_\rho^1 u_k(r) \frac{R_k(r,\rho)}{\sqrt{r-\rho}} \, dr = v_k(\rho) \; ,$$

where

$$(5) \qquad R_k(r,\rho) = \sqrt{r-\rho} \sum_{j=1}^{2} \sqrt{1+(r \frac{\partial \varphi_j}{\partial r})^2} \, \exp[ik(-1)^j \psi_j(r,\rho) \sqrt{r-\rho}].$$

The functions $u_k(r)$ and $v_k(\rho)$ in (4) are complex-valued functions of a real variable.

Using the representation (1) for the functions φ_j, we can easily derive the formula

$$(6) \qquad R_k(r,\rho) = \sum_{j=1}^{2} \sqrt{r-\rho+r^2[\tfrac{1}{2}\psi_j(r,\rho) + (r-\rho) \tfrac{\partial}{\partial r} \psi_j(r,\rho)]^2}$$

$$\cdot \exp[ik(-1)^j \psi_j(r,\rho) \sqrt{r-\rho} \;] \; .$$

It is apparent from this that $R_k(r,\rho)$ is continuous in the domain $D\{0 < \rho \leq r \leq 1\}$ together with $\sqrt{r-\rho} \; \frac{\partial}{\partial \rho} R_k(r,\rho)$. In addition

$$(6a) \qquad R_k(\rho,\rho) = \rho\psi_1(\rho,\rho) \neq 0, \qquad\qquad (0<\rho\leq1) \; .$$

These properties can be used to easily reduce the integral equation (4) to a VOLTERRA equation of second kind. Indeed, apply to (4) the operator L defined by

$$(7) \qquad Lv = \int_s^1 (\rho-s)^{-\frac{1}{2}} v(\rho) d\rho$$

and then change the order of integration. This yields

$$(8) \qquad \int_s^1 u_k(r)Q_k(r,s)dr = Lv_k,$$

where

$$(9) \qquad Q_k(r,s) = \int_s^r \frac{R_k(r,\rho)}{\sqrt{(r-\rho)(\rho-s)}} \, d\rho$$

Making the change of variables

$$(9a) \qquad \rho = \frac{r+s}{2} + \frac{r-s}{2} \cos \theta$$

in the last integral, we arrive at

$$(10) \qquad Q_k(r,s) = \int_0^\pi R_k(r, \tfrac{r+s}{2} + \tfrac{r-s}{2} \cos\theta) \, d\theta \ .$$

Formula (10) implies that $Q_k(r,s)$ is a continuous function of r and s in D together with $\sqrt{r-s} \, \frac{\partial}{\partial s} Q_k(r,s)$. In addition,

$$(10a) \qquad Q_k(s,s) = \pi s \psi_1(s,s) \ .$$

Differentiating equation (8) with respect to s and then dividing both sides of the equation by $Q_k(s,s)$, we obtain

$$(11) \qquad -u_k(s) + \int_s^1 u_k(r) \, T_k(r,s) dr = \frac{1}{Q_k(s,s)} \, \frac{\partial}{\partial s} L v_k,$$

where

$$(12) \qquad T_k(r,s) = \frac{1}{Q_k(s,s)} \, \frac{\partial}{\partial s} Q_k(r,s) \ .$$

The kernel of equation (11) has a weak polar singularity. At the same time, $\sqrt{r-s} \, T_k(r,s)$ is continuous for $0 < s \le r \le 1$ and hence (11) has a unique continuous solution $u_k(s)$. Recalling that the $u_k(r)$ are the FOURIER coefficients of $u(r, \varphi)$, we conclude that a solution to equation (2) is also unique.

Since the kernels of equation (11) are known for the family of curves in question, namely, they are given by (12), a resolvent may be constructed for each such equation. As the result of inverting (11), we obtain the set of formulas,

$$(13) \qquad u_k(r) = M_k v,$$

$$(k = 0, \pm 1, \pm 2, \ldots)$$

wherein the M_k are fixed operators for the family of curves in question.

We now consider what properties the functions $v(\rho,\alpha)$ must have as images of the functions $u(r, \varphi)$ under the mapping defined by (2). As before, we make a slight contraction in the class of continuous functions. Namely, we consider functions $u(r, \varphi)$ having absolutely convergent FOURIER series, or more precisely, for which

$$(14) \qquad \sum_{k=-\infty}^\infty \max_r |u_k(r)| < \infty \ .$$

The set of functions $u(r, \varphi)$ for which inequality (14) holds will be
labeled U. The set of images $v(\rho, \alpha)$ of U under the mapping (2) will
be denoted by V.

Theorem 6: *The set V possesses the following properties:*

1°. For each $v(\rho, \alpha)$, the functions $M_k v$ $(k = 0, \pm 1, \pm 2, \ldots)$ are
continuous.

2°. For each $v(\rho, \alpha)$,

$$(15) \qquad \sum_{k=-\infty}^{\infty} \max_r |M_k v| = \alpha_v < \infty \quad .$$

3°. Each $v \in V$ admits the estimate

$$(16) \qquad |v(\rho, \alpha)| \leq \alpha_v \cdot \sum_{j=1}^{2} \int_{\rho}^{1} \sqrt{1 + (r \frac{\partial \varphi_j}{\partial r})^2} \, dr \quad .$$

The fulfillment of the first and second conditions follows from (13)
and (14). To prove the validity of inequality (16), we consider the
inversion formula for the given problem. After the FOURIER coefficients
of $u(r, \varphi)$ have been found using (13), $u(r, \varphi)$, being a member of U,
can be determined through the formula

$$(17) \qquad u(r, \varphi) = \sum_{k=-\infty}^{\infty} (M_k v) e^{ik\varphi} \quad .$$

The substitution of this expression into formula (2) must reduce it to
an identity and so we have

$$(18) \qquad \sum_{j=1}^{2} \int_{\rho}^{1} \sum_{k=-\infty}^{\infty} (M_k v) e^{ik\varphi} \sqrt{1 + (r \frac{\partial \varphi_j}{\partial r})^2} \, dr = v(\rho, \alpha) \quad .$$

This identity leads to inequality (16) in a trivial way and so all three
conditions have been proved. The fact that $v(\rho, \alpha)$ belongs to V is
also a sufficient condition for a solution to exist for equation (2).
Indeed, suppose $v \in V$. From it, construct the series

$$(18a) \qquad \sum_{k=-\infty}^{\infty} (M_k v) e^{ik\varphi} \quad .$$

By conditions 1° and 2° the series converges to a continuous
function $u(r, \varphi)$ for which the functions $M_k v$ are its FOURIER coefficients. Hence by condition 2° it belongs to U.

On the basis of $u(r, \varphi)$, we can compute function $\tilde{v}(\rho, \alpha)$ defined by

$$(19) \quad \sum_{j=1}^{2} \int_{\rho}^{1} \sum_{k=-\infty}^{\infty} (M_k v) e^{ik\varphi} \sqrt{1 + (r \frac{\partial \varphi_1}{\partial r})^2} \, dr = \tilde{v}(\rho, \alpha) \quad .$$

Since $u(r, \varphi) \in U$, it follows that $\tilde{v} \in V$, and we can apply M_k to (19). By formula (13) we then obtain

$$(19a) \qquad M_k v = M_k \tilde{v} \quad , (k=0, \pm 1, \pm 2, \ldots) \quad .$$
$$w = v - \tilde{v}$$

Or by the linearity of the operator M_k,

$$(19b) \qquad M_k w = 0, \qquad (k=0, \pm 1, \pm 2, \ldots)$$

The element w belongs to V and it therefore satisfies inequality (16). This implies that $w(\rho, \alpha) \equiv 0$, or in other words, $\tilde{v}(\rho, \alpha) = v(\rho, \alpha)$. But this means that to each $v(\rho, \alpha) \in V$ we have constructed a solution to equation (2). By Theorem 5, it is unique.

Collecting the above results, we arrive at the following theorem:

Theorem 7: *A necessary and sufficient condition for a solution to exist to equation (2) belonging to U is that v(ρ,α) belong to V.*

Remark. We have considered the case where the entire unit disc is covered by curves satisfying conditions 1°-3°. If we consider only curves in a subdomain D of the unit disc bounded by a curve of the family and the unit circle, then by the local convergence of FOURIER series, we can construct a solution to (2) in D in a similar way.

4. On the Problem of Determining a Function from its Mean Values over Circles

In COURANT'S book [6], the question of determining an even function of s, $u(x_1, x_2, \ldots x_n, s)$ from its mean values over spheres of arbitrary radius with centers at points of the hyperplane $s = 0$ is considered. It is shown that such a function is unique within the class of continuous functions. At the same time, an algorithm is given for constructing the function approximately. However, up until now, no inversion formula has been developed for the problem and the question of existence of a solution has gone unclarified. In what follows, we shall in a certain sense fill in this gap. But in order to simplify the computations, we shall consider a planar version of the problem in terms of circles even though it has no basic significance.

Thus, suppose that the integrals of $u(x,s)$ are known along circles centered on the line $s = 0$, i.e.

$$(1) \qquad \int_0^{2\pi} u(x+r\cdot\cos\varphi,\ r\cdot\sin\varphi)d\varphi = v(x,r) \quad .$$

Applying to (1) the operator L defined by

$$(2) \qquad Lv = xv(x,r) + r\frac{\partial}{\partial x}\int_0^r \rho v(x,\rho)d\rho$$

we obtain

$$Lv = x\int_0^{2\pi} u(x+r\cdot\cos\varphi,r\cdot\sin\varphi)d\varphi + r\frac{\partial}{\partial x}\iint_{(x-\xi)^2+s^2\leq r^2} u(\xi,s)d\xi ds$$

$$(2a) \quad = x\int_0^{2\pi} u(x+r\cdot\cos\varphi,r\cdot\sin\varphi)d\varphi + r\frac{\partial}{\partial x}\int_{x-r}^{x+r} d\xi\int_{-\sqrt{r^2-(\xi-x)^2}}^{\sqrt{r^2-(\xi-x)^2}} \frac{u(\xi,s)}{\sqrt{r^2-(\xi-x)^2}}\ ds$$

$$= x\int_0^{2\pi} u(x+r\cdot\cos\varphi,r\cdot\sin\varphi)d\varphi + \int_0^{2\pi} u(x+r\cdot\cos\varphi,r\cdot\sin\varphi)r\cdot\cos\varphi d\varphi$$

$$= \int_0^{2\pi} u(x+r\cdot\cos\varphi,r\cdot\sin\varphi)(x+r\cdot\cos\varphi)\ d\varphi\ .$$

Let $L^k v$ denote the k-fold application of L to $v(x,r)$. Then in a similar way, we have

$$(3) \quad L^k v = \int_0^{2\pi} u(x+r\cdot\cos\varphi,\ r\cdot\sin\varphi)(x+r\cdot\cos\varphi)^k d\varphi \quad .$$

In this connection, $L^0 v = v(x,r)$.

From this it will follow in particular that the even function (of s) $u(x,s)$ can be determined from $v(x,r)$ in unique fashion. In analogy with Sec. 1, we let M_k be an operator defined by

$$(4) \qquad M_k v = \frac{1}{\pi}\sum_{j=0}^{[\frac{k}{2}]} (-1)^j \left(\sum_{\ell=0}^{[\frac{k}{2}]-j} \binom{k}{2(j+\ell)}\binom{j+\ell}{\ell}\right) r^{2j-k} L^{k-2j} v,$$

$$(k = 1,2,\ldots)$$

$$M_0 v = \frac{1}{2\pi}v(x,r).$$

Its application to equation (1) results in the following set of
relations:

$$(5) \quad M_k v = \frac{1}{2\pi} \int_0^{2\pi} u(x + r \cdot \cos\varphi, \ r \cdot \sin\varphi) \left\{ \left[t + \sqrt{t^2 - 1} \right]^k + \left[t - \sqrt{t^2 - 1} \right]^k \right\}$$

$$d\varphi \ (t = \tfrac{x}{r} + \cos\varphi) \qquad (k = 1, 2, \ldots)$$

For x = 0, they may be written in the form

$$(6) \quad [M_k v]_{x=0} = \frac{1}{\pi} \int_0^{2\pi} u(r \cdot \cos\varphi, \ r \cdot \sin\varphi) \cos k\varphi \, d\varphi \ .$$

$$(k = 1, 2, 3, \ldots)$$

From this we obtain an inversion formula for continuous functions belonging to U, namely,

$$(7) \quad u(r \cdot \cos\varphi, \ r \cdot \sin\varphi) = \sum_{k=0}^{\infty} [M_k v]_{x=0} \cdot \cos k\varphi \ .$$

Formula (1) now easily leads to an equality for the function v(x,r)
corresponding to which the solution to (1) is given by (7). It amounts
to the following:

$$(8) \quad [v(x,r)] \leq 2\pi \sum_{k=0}^{\infty} \max_r [M_k v]_{x=0} \ .$$

Finally, it is possible to formulate a theorem in terms of the functions
$M_k v$ as was done previously.

Theorem 8: *A necessary and sufficient condition for a solution to exist
for equation (1) belonging to the set U is that v(x,r)
belong to the set V of functions satisfying the following
conditions :*

*1. For each $v(x,r) \in V$, the functions $M_k v$ (k = 0, 1, ...)
are continuous.*

2. $\sum_{k=0}^{\infty} \max_r |M_k v|_{x=0} < \infty$.

3. Each $v(x,r) \in V$ satisfies inequality (8).

The proof of the theorem is carried out along the same lines as before
and so we shall not stop to give it.

CHAPTER 2

Linearized Inverse Dynamic Problem for the
Telegraph Equation

Consider the problem of determining the coefficient $a(x,y,z)$ in the telegraph equation

$$\frac{\partial^2 u}{\partial t^2} = \frac{\partial^2 u}{\partial x^2} + \frac{\partial^2 u}{\partial y^2} + \frac{\partial^2 u}{\partial z^2} + a(x,y,z)u$$

in the domain $z \geq 0$ from the value of the solution to the CAUCHY problem for the equation at $z = 0$. The CAUCHY data may in principle depend on several parameters.

The first one to examine this sort of problem was Ju.M. BEREZANSKIĬ [4]. He showed that $a(x,y,z)$ can be determined uniquely in the class of piecewise analytic functions from the function $u(x,y,0,x_o,y_o,t)$. Here x_o and y_o are parameters connected with the CAUCHY data that take on values in a certain two-dimensional region. Thus, to reconstruct a three-dimensional function, it is necessary to know a function of five variables.

In this chapter, we shall examine two kinds of linear approximations to the problem. Each linearized problem is then reduced to an integral-geometric problem considered in Chapter 1. To determine the coefficient, we need to know information of the same dimensionality as that of the coefficient $a(x,y,z)$.

1. Statement of the Inverse Problem and Its Linearization

Consider the telegraph equation

$$(1) \qquad \frac{\partial^2 u}{\partial t^2} = \Delta u + a(M)u + f(M,M_o,t)$$

for the function $u(M,M_o,t)$ in the domain $z \geq 0$ under the following initial and boundary conditions:

$$(2) \qquad u(M,M_o,0) = \frac{\partial}{\partial t} u(M,M_o,0) = 0,$$

$$(3) \qquad \frac{\partial}{\partial z} u(M_1,M_o,t) = 0, \qquad t > 0 \quad .$$

Here $M(x,y,z)$, $M_o(x_o,y_o,0)$ and $M_1(x,y,0)$ are points in three-dimensional space, Δ is the Laplacian in x,y and z, and $a(M)$ is a continuous function in the domain $z \geq 0$. Throughout the following we shall assume that $f(M,M_o,t)$ is a function of the form

$$(4) \qquad\qquad f(M,M_o,t) = \delta(M-M_o,t)$$

where $\delta(M-M_o,t)$ is the DIRAC delta-function (see [11]). By a solution $u(M,M_o,t)$ to equation (1) under conditions (2) and (3), we shall mean a generalized solution of the equation regular at infinity. If $a(M)$ is given, then determining the function $u(M,M_o,t)$ is called the direct problem for equation (1). We now state the corresponding inverse problem: it is required to find continuous function $a(M)$ knowing the value at $z = 0$ of the solution to equation (1) under conditions (2) and (3).

Thus suppose that the function

$$(5) \qquad\qquad \varphi(M_1,M_o,t) = u(M_1,M_o,t)$$

has been prescribed.

It is convenient to reduce the stated problem by use of the boundary condition (3) to an equivalent problem for the whole space. This we do by continuing $a(M)$ and $f(M,M_o,t)$ as even functions into the half-space $z < 0$.

We represent the solution $u(M,M_o,t)$ in the form

$$(6) \qquad\qquad u(M,M_o,t) = u_o(M,M_o,t) + u_1(M,M_o,t)$$

$u_o(M,M_o,t)$ being a function satisfying the equation

$$(7) \qquad\qquad \frac{\partial^2 u}{\partial t^2} = \Delta u + \delta(M-M_o,t)$$

and boundary conditions (2) and (3). We then obtain for $u_1(M,M_o,t)$ the equation

$$(8) \qquad\qquad \frac{\partial^2 u_1}{\partial t^2} = \Delta u_1 + a(M)\left[u_o + u_1\right]$$

and conditions similar to (2) and (3). If $a(M)$ is allowed to decrease indefinitely in this equation, its solution $u_1(M,M_o,t)$ will obviously tend to zero. Taking this into consideration, we shall regard a(M) to be so small that the second order term in (8) may be neglected, namely, the product $a(M)u_1(M,M_o,t)$. As a result, we obtain the linearized equation

$$(9) \qquad \frac{\partial^2 u_1}{\partial t^2} = \Delta u_1 + a(M)u_o(M,M_o,t)$$

and it is this equation for which we shall consider the problem of re-constructing a(M) from the known value of its solution for z = 0,

$$(10) \qquad u_1\Big|_{z=0} = \varphi_1(M_1,M_o,t)$$

under conditions on $u_1(M,M_o,t)$ analogous to (2) and (3).

2. Linearized One-Dimensional Inverse Problem in Two-Dimensional Space

In this section, we shall deal with the linearized problem of deter-mining a(M) in the half-plane $y \geq 0$ on the assumption it is just a function of y. Carrying over the discussion of Sec. 1 to the two-dimensional case, we arrive at the problem of determining a(y) from the solution to the equation

$$(1) \qquad \frac{\partial^2 u_1}{\partial t^2} = \frac{\partial^2 u_1}{\partial x^2} + \frac{\partial^2 u_1}{\partial y^2} + a(y)u_o(x,y,t)$$

in the domain $y \geq 0$ under the initial and boundary conditions

$$(2) \qquad u_1(x,y,0) = \frac{\partial}{\partial t} u_1(x,y,0) = 0,$$

$$(3) \qquad \frac{\partial}{\partial y} u_1(x,0,t) = 0, \qquad (t > 0)$$

the solution being known at a single point:

$$(4) \qquad u_1(x_1,0,t) = \varphi(t) \quad .$$

The point M_o will be regarded as the origin. The solution $u_o(x,y,t)$ of the two-dimensional analogue of equation (7) of Sec. 1 in the half-plane $y \geq 0$ is given by [32]

$$(5) \qquad u_o(x,y,t) = \begin{cases} \dfrac{1}{\pi\sqrt{t^2-x^2-y^2}}, & t^2 \geq x^2 + y^2 \\[2mm] 0, & t^2 < x^2 + y^2 \end{cases}$$

while the solution to (1) under conditions (2) and (3) can be expressed
as

$$(6) \qquad u_1(x,y,t) = \frac{1}{2\pi} \int_0^t \left[\iint_{r \leq t-\tau} \frac{a(\eta)u_0(\xi,\eta,\tau)d\xi d\eta}{\sqrt{(t-\tau)^2 - r^2}} \right] d\tau$$

wherein

$$(6a) \qquad r = \sqrt{(x-\xi)^2 + (y-\eta)^2} .$$

We now let the point (x,y) tend to $(x_1,0)$ in formula (6). This
results in the following integral equation of the first kind for the
function $a(y)$:

$$(7) \qquad \varphi(t) = \frac{1}{2\pi} \int_0^t \left[\iint_{r_1 \leq t-\tau} \frac{a(\eta)u_0(\xi,\eta,\tau)d\xi d\eta}{\sqrt{(t-\tau)^2 - r_1^2}} \right] d\tau$$

$$(r_1 = \sqrt{(x_1-\xi)^2 + \eta^2})$$

The region of integration in (7) is the interior of the cone

$$(8) \qquad \tau = t - \sqrt{(\xi-x_1)^2 + \eta^2}$$

and as is apparent from (5), the function $u_0(x,y,t)$ is nonvanishing
inside the cone

$$(9) \qquad \tau = \sqrt{\xi^2 + \eta^2} .$$

Hence, the integration in (7) actually extends over the region common
to the two cones.

Introduce the notation

$$(10) \qquad \begin{cases} r_2(\xi,\eta) = \sqrt{\xi^2 + \eta^2} , \\[2mm] \eta_1 = \frac{1}{2}\sqrt{t^2 - x_1^2} , \\[2mm] \xi_1 = \frac{x_1}{2} - \frac{t}{2\eta_1}\sqrt{\eta_1^2 - \eta^2} , \\[2mm] \xi_2 = \frac{x_1}{2} + \frac{t}{2\eta_1}\sqrt{\eta_1^2 - \eta^2} . \end{cases}$$

We change the order of integration in (7) and take into account the
evenness of the integrand in η. Equation (7) can thus be rewritten as

(11)
$$\varphi(t) = \int_0^{n_1(t)} a(n) \, K(t,n) \, dn \quad ,$$

where

(12)
$$K(t,n) = \frac{1}{\pi^2} \int_{\xi_1}^{\xi_2} \int_{r_2}^{t-r_1} \frac{d\tau \, d\xi}{\sqrt{[(t-\tau)^2 - r_1^2][\tau^2 - r_2^2]}} \quad .$$

We make a change of variables in the iterated integral (12) setting

(12a)
$$z = \frac{2\tau - (t - r_1 + r_2)}{t - r_1 - r_2} \quad , \quad u = \frac{n_1}{t} \frac{2\xi - x_1}{\sqrt{n_1^2 - n^2}} \quad .$$

The formula for the kernel $K(t,n)$ then becomes

(13)
$$K(t,n) = \frac{t}{\pi^2 n_1^2} \sqrt{n_1^2 - n^2} \int_{-1}^{1} \int_{-1}^{1} \frac{dz \, du}{\sqrt{(1-z^2)[4\rho_1 + (t-\rho_1-\rho_2)(1-z)][4\rho_2 + (t-\rho_1-\rho_2)(1+z)]}} \quad .$$

Here we have introduced the notation

(14)
$$\begin{cases} \rho_1 = r_1(\frac{x_1}{2} + u \frac{t}{2n_1} \sqrt{n_1^2 - n^2}, \, n) \\[2mm] \rho_2 = r_2(\frac{x_1}{2} + u \frac{t}{2n_1} \sqrt{n_1^2 - n^2}, \, n) \end{cases}$$

Equation (11) is a VOLTERRA equation of first kind. To answer the question as to whether it is reducible to an equation of second kind, we must ascertain the behavior of its kernel as $n \to n_1(t)$. We shall now do this. Using (10) and (14), we can easily deduce the following relations for $n \to n_1$:

(14a)
$$\rho_1 = \frac{t}{2} - \frac{x_1}{2n_1} \sqrt{n_1^2 - n^2} - \frac{1}{t}(n_1^2 - n^2)(u^2 - 1) + O_1((n_1 - n)^{\frac{3}{2}}),$$

$$\rho_2 = \frac{t}{2} + \frac{x_1}{2n_1} \sqrt{n_1^2 - n^2} - \frac{1}{t}(n_1^2 - n^2)(u^2 - 1) + O_2((n_1 - n)^{\frac{3}{2}})$$

where $O_1((n_1 - n)^{\frac{3}{2}})$ and $O_2((n_1 - n)^{\frac{3}{2}})$ are infinitesimals of order $(n_1 - n)^{\frac{3}{2}}$. Hence we conclude that the kernel

(15)
$$K(t,n) = \frac{t}{\pi^2 n_1^2} \sqrt{n_1^2 - n^2} \int_{-1}^{1} \int_{-1}^{1} \frac{1}{\sqrt{1 - z^2}} \left\{ \frac{1}{2t} + O_3((n_1 - n)) \right\} dz du$$

$$= \sqrt{n_1 - n} \left[\frac{\sqrt{2}}{\pi \sqrt{n_1}} + O(n_1 - n) \right]$$

as $n \to n_1$.

This formula remains valid as $\eta \to 0$. A more detailed analysis shows that the origin is a singular point for $K(t,\eta)$, namely, its limit there does not exist. However $K(t,\eta)$ remains bounded in a neighborhood of the origin. Formula (15) leads to

$$(15a) \qquad K(t,\eta) = \sqrt{\eta_1 - \eta} \; K_1(t,\eta) \; .$$

The function $K_1(t,\eta)$ is continuous together with its derivatives except at the origin.

We now replace the variable t in (11) by its value in terms of η_1. As a result, equation (11) becomes

$$(16) \qquad \varphi(\sqrt{x_1^2 + 4\eta_1^2}) = \int_0^{\eta_1} a(\eta) \; K(\sqrt{x_1^2 + 4\eta_1^2}, \eta) d\eta \; .$$

Apply to the left-hand and right-hand sides of (16) the operator L defined by

$$(17) \qquad L\varphi = \frac{d}{ds} \int_0^s \sqrt{\frac{2s}{s-\eta_1}} \frac{d\varphi}{d\eta_1} d\eta_1 \; .$$

We thus obtain

$$L\varphi = \frac{d}{ds} \int_0^s \sqrt{\frac{2s}{s-\eta_1}} \left[\int_0^{\eta_1} a(\eta) \frac{\partial}{\partial \eta_1} K(\sqrt{x_1^2 + 4\eta_1^2}, \eta) d\eta \right] d\eta_1$$

$$(17a) \qquad = \frac{d}{ds} \int_0^s a(\eta) \left[\int_\eta^s \sqrt{\frac{2s}{s-\eta_1}} \frac{\partial}{\partial \eta_1} K(\sqrt{x_1^2 + 4\eta_1^2}, \eta) d\eta_1 \right] d\eta$$

$$= a(s) \; R(s,s) + \int_0^s a(\eta) \frac{\partial}{\partial s} R(s,\eta) \; d\eta \; ,$$

where

$$(18) \qquad R(s,\eta) = \int_\eta^s \sqrt{\frac{2s}{s-\eta_1}} \frac{\partial}{\partial \eta_1} K(\sqrt{x_1^2 + 4\eta_1^2}, \eta) d\eta_1 \; .$$

From (18) it is easy to deduce that

$$(18a) \qquad R(s,s) = 1 \; .$$

As a result, we arrive at the following VOLTERRA equation of second kind

$$(19) \qquad a(s) + \int_0^s a(\eta) \frac{\partial}{\partial s} R(s,\eta) d\eta = L\varphi \; .$$

The kernel of this equation is continuous everywhere with the exception
of the point $s = 0$. It is possible to show further that the function
$s\frac{\partial}{\partial s} R(s,\eta)$ remains continuous at $s = 0$ and that there exists an s^{*}
such that

$$(19a) \qquad\qquad |s \frac{\partial}{\partial s} R(s,\eta)| \leq \alpha < 1$$

in the domain $\{0 \leq s < s^{*}, \eta \leq s\}$.

From this it follows that the determination of $a(s)$ from the function
$\varphi(t)$ is unique. On the other hand, a solution will exist if and only
if $\varphi(t)$ is such that $L\varphi$ exists and is continuous.

3. Two Formulations of the Linearized Inverse Problem in Three-Dimensional Space

For three-dimensional space, the function $u_o(M,M_o,t)$ that satisfies
equation (7) of Sec. 1 and homogeneous initial and boundary conditions
such as (2) and (3) is given by

$$(1) \qquad\qquad u_o(M,M_o,t) = \frac{\delta(t-r(M,M_o))}{2\pi\, r(M,M_o)} \qquad .$$

In this formula $r(M,M_o)$ is the distance between the points M and M_o.

The solution to equation (9) of Sec. 1 under corresponding initial and
boundary conditions is given by KIRCHHOFF'S formula :

$$(2) \qquad u_1(M,M_o,t) = \frac{1}{4\pi} \iiint\limits_{r(P,M)\leq t} \frac{a(P)\, u_o(P,M_o,t-r(P,M))}{r(P,M)}\, dv_P \qquad .$$

The integration variable $P(\xi,\eta,\zeta)$ runs over the points of a ball of
radius t centered at the point M.
With (2) as point of departure, one may consider two formulations of
the linearized inverse problem.

First Formulation: Let $M_o(x_o,y_o,0)$ be a variable point and let
$u_1(M_o,M_o,t) = \psi(M_o,t)$ be a given function. Letting the point M
approach M_o in formula (2) and using expression (1), we obtain

$$(3) \qquad\qquad \psi(M_o,2t) = \frac{1}{8\pi^2 t^2} \iint\limits_{S_{M_o,t}} a(P)dS$$

where $S_{M_o,t}$ is a sphere of radius t with center at M_o and dS is spherical surface element.

Thus in this version of the linearized inverse problem we arrive at the integral-geometric problem of determining a function from its mean values over spheres of arbitrary radius with centers at points of the plane $z = 0$. This problem, as we have already pointed out, has a unique solution and thus we have the following theorem.

Theorem 9: *The coefficient* $a(M)$ *in the linearized inverse problem can be determined from* $u_1(M_o, M_o, t)$ *in a unique manner.*

We mention that the results in COURANT'S book [6] show that it suffices for the center M_o to belong to a point set in the plane $z = 0$ lying in an ε-neighborhood of some fixed point of this plane.

In the case where $a(M) = a(z)$, it suffices to know the function $u(M_o, M_o, t)$ at a fixed point M_o. The function $a(z)$ can then be found explicitly. To this end, introduce spherical coordinates at the point M_o, letting θ be the angle between the z-axis and the radius vector of the variable point of integration. Formula (3) can then be written in the form

$$(4) \qquad \psi(M_o, 2t) = \frac{1}{4\pi} \int_0^\pi a(t \cos\theta) \sin\theta d\theta \ .$$

Equation (4) has the explicit solution

$$(5) \qquad a(z) = 2\pi \frac{d}{dz} [z\psi(M_o, 2z)] \ .$$

Second Formulation: The point $M_o(x_o, y_o, 0)$ is fixed and the function $u_1(M_1, M_o, t) = (M_1, t)$ is given, where $M_1(x, y, 0)$ is a variable point of the plane $z = 0$.

Substituting the expression for $u_o(M, M_o, t)$ given by (1) into (2), we find in this case that the function $u_1(M, M_o, t)$ is given by

$$(6) \qquad u_1(M, M_o, t) = \frac{1}{8\pi^2} \iiint_{r(P,M) \leq t} \frac{a(P)\delta(t - r(P,M) - r(P,M_o))}{r(P,M)r(P,M_o)} dv_P \ .$$

It is apparent from this formula that the integrand is nonvanishing only on a surface whose equation is

(7) $$r(P,M_o) + r(P,M) = t \quad .$$

This is nothing more than the equation of an ellipsoid of revolution with foci at the points M_o and M. Since M_o is fixed, it may be regarded as the origin of our fixed cartesian coordinate system. We also introduce a moving cartesian coordinate system ξ', η', ζ' with ζ'-axis coinciding with the line passing through the foci M and M_o of the ellipsoid. To simplify the integral in (6), we consider in addition to the ellipsoid (7) a family of confocal ellipsoids of revolution given by

(8) $$r(P,M_o) + r(P,M) = \tau \quad .$$

We now pass from cartesian coordinates ξ', η', ζ' to spherical coordinates r, θ, φ by means of

(8a)
$$\begin{cases} \xi' = r \sin\theta \cos\varphi, \\ \eta' = r \sin\theta \sin\varphi, \\ \zeta' = r \cos\theta \quad . \end{cases}$$

The ellipsoids (8) may then be written in the form

(9) $$r = \frac{1}{2} \frac{\tau^2 - \rho^2}{\tau - \rho \cos\theta} \quad ,$$

where ρ is the distance between M_o and M.

It is not hard to show that

(9a) $$\frac{dv_P}{r(P,M)r(P,M_o)} = \frac{2r^2}{\tau^2 - \rho^2} \sin\theta \; d\theta \; d\varphi \; d\tau$$

and so expression (6) can be represented in the form

(10) $$u_1(M,M_o,t) = \frac{1}{4\pi^2[t^2 - r^2(M,M_o)]} \iint_{S_{M,t}} a(P)r^2(P,M_o)d\omega \quad ,$$

where $S_{M,t}$ denotes the ellipsoid of revolution defined by (7) and $d\omega$ is the element of solid angle with vertex at point M_o.

Letting the point M tend to $M_1(x,y,0)$ in relation (10), we obtain an integral equation of first kind for a(M), namely,

(11) $\iint\limits_{S_{M_1,t}} a(P)r^2(P,M_o)d\omega = 4\pi^2[t^2-r^2(M_1,M_o)]\,\varphi(M_1,t)$.

Since the factor $r^2(P,M_o)$ can always be grouped with the function $a(P)$ and their product satisfies a HÖLDER condition at M_o, we can apply the results of Sec. 1 of Chapt. 1 to arrive at the following theorem.

Theorem 10: *The coefficient* $a(M)$ *in the linearised telegraph equation can be determined from the function* $u_1(M_1,M_o,t)$ *in a unique manner.*

Finally, the algorithm of Sec. 1, Chapt. 1 may be used to construct $a(M)$.

4. Derivation of a Nonlinear Differential Equation for the Inverse Problem

We next consider how the coefficient $a(M)$ in the telegraph equation is related to the solution itself. To this end, we shall make use of equation (8) of the first section of this chapter. If we do not neglect the term $a(M)u_1$ in this equation and use similar reasoning to that of the preceding section, we can derive the formula

$$(1) \quad u_1(M,M_o,t) = \frac{1}{4\pi^2[t^2-r^2(M,M_o)]} \iint\limits_{S_{M,t}} a(P)r^2(P,M_o)d\omega$$

$$+ \frac{1}{4\pi} \iiint\limits_{D_{M,t}} \frac{a(P)u_1(P,M_o,t-r(P,M))}{r(P,M)} dv_P \quad .$$

$D_{M,t}$ here denotes the domain having the ellipsoid of revolution $S_{M,t}$ as boundary. It is apparent from this relation that the solution $u_1(M,M_o,t)$ is bounded. Therefore, if we pass to the limit in (1) letting $t \to r(M,M_o)$, the integral over $D_{M,t}$ will vanish. Moreover, the ellipsoid of revolution degenerates to the line segment joining M and M_o. Denoting by $r^o(M_o,M)$ the unit vector in the direction of the line from M_o to M and letting the parameter t tend to $r(M,M_o)$ in (1), we obtain in the limit the expression

$$(2) \quad u_1(M,M_o,r(M,M_o)) = \frac{1}{4\pi r(M,M_o)} \int\limits_0^{r(M,M_o)} a(r \cdot \vec{r}^o(M_o,M))dr \quad .$$

Computing the directional derivative of both members of this relation in the direction $\vec{r}^o(M_o,M)$, we conclude that

$$(3) \qquad a(M) = 4\pi \frac{\partial}{\partial \vec{r}^o(M_o,M)} [r(M,M_o)u_1(M,M_o,r(M,M_o))] \quad .$$

Formula (3) which we have obtained for the coefficient in the telegraph equation has the same type of structure as the formula arising in the spectral version of the one-dimensional inverse STURM-LIOUVILLE problem (see [1] and [13]).

Substituting this expression for $a(M)$ in formula (8) of Sec. 1, we arrive at the following nonlinear differential equation with shifted argument:

$$(4) \qquad \frac{\partial^2 u_1}{\partial t^2} = \Delta u_1 + 4\pi(u_o+u_1) \frac{\partial}{\partial \vec{r}^o(M_o,M)} [r(M_o,M)u_1(M,M_o,r(M,M_o))] \quad .$$

If this equation should be solvable under the initial conditions

$$(5) \qquad u_1|_{t=0} = \frac{\partial}{\partial t} u_1|_{t=0} = 0$$

and boundary conditions

$$(6) \qquad \left. \begin{aligned} u_1|_{z=0} &= \varphi(M_1,t) \ , \\ \frac{\partial u_1}{\partial z}\bigg|_{z=0} &= 0 \end{aligned} \right\}$$

then we could afterwards find $a(M)$ using (3). The questions of existence and uniqueness of a solution to the above problem for (4) require further investigation.

CHAPTER 3

Linearized Inverse Kinematic Problem for the Wave Equation

1. Formulation of the Problem and Its Linearization

Consider the wave equation

$$(1) \qquad n^2 \frac{\partial^2 u}{\partial t^2} = \Delta u ,$$

where n is a function of x and y and Δ is the LAPLACIAN with respect to the same variables. Let $u(x,x_o,y,t)$ be the generalized solution to equation (1) in the half-plane $y > 0$ satisfying the following initial and boundary conditions:

$$(2) \qquad u(x,x_o,y,t)\big|_{t=0} = \frac{\partial}{\partial t} u(x,x_o,y,t)\big|_{t=0} = 0 ,$$

$$\frac{\partial}{\partial y} u(x,x_o,y,t)\big|_{y=0} = \delta(x-x_o,t) ,$$

$\delta(x-x_o,t)$ being the DIRAC delta-function. Further, let $\tau(x_1,x_o)$ be the maximum of the numbers τ such that for given x_o and x_1, the support of $u(x_1,x_o,0,t)$ is a subset of the interval $\tau \leq t < \infty.$ *

We pose the following inverse problem for equation (1): Given the function $\tau(x_1,x_o)$, to determine $n(x,y)$. This problem is called the inverse kinematic problem. It was studied in the papers [15] and [37] for the case where the function n is independent of x. Some results for the problem are to be found in [3].

The function $\tau(x_1,x_o)$ is the minimum of the functional

$$(3) \qquad J(\Gamma) = \int_{\Gamma(x_1,x_o)} n(x,y)ds$$

where $\Gamma(x_1,x_o)$ is a curve joining the points $(x_1,0)$ and $(x_o,0)$ lying in the upper half-plane and ds is the element of arclength along

---

*In the process characterized by (1), $\tau(x_1,x_o)$ is the time at which a disturbance produced at $(x_o,0)$ reaches $(x_1,0)$.

the curve. Now let

$$(4) \qquad n(x,y) = n_0(x,y) + n_1(x,y)$$

where $n_0(x,y)$ is a given function and $n_1(x,y)$ is a sufficiently smooth small function. Correspondingly, $\tau(x_1,x_0)$ may be represented as the sum

$$(5) \qquad \tau(x_1,x_0) = \tau_0(x_1,x_0) + \tau_1(x_1,x_0) \quad .$$

Here $\tau_0(x_1,x_0)$ corresponds to the function $n_0(x,y)$, or in other words, is a minimum of the functional

$$(6) \qquad J_0(\Gamma) = \int_{\Gamma(x_1,x_0)} n_0(x,y)ds \ .$$

The function $\tau_1(x_1,x_0)$ is then also small. The following theorem holds :

Theorem 11: *The function $\tau_1(x_1,x_0)$ can be represented to within in-finitesimals of order n_1^2 by*

$$(7) \qquad \tau_1(x_1,x_0) = \int_{\Gamma^0(x_1,x_0)} n_1(x,y)ds$$

where $\Gamma^0(x_1,x_0)$ is the curve of the family $\Gamma(x_1,x_0)$ for which the minimum of the functional (6) is attained.

Denote by $\Gamma^1(x_1,x_0)$ the curve of the family $\Gamma(x_1,x_0)$ for which the minimum of the functional (3) is attained. We can then write down the following relation:

$$(8) \qquad \tau(x_1,x_0) - \int_{\Gamma^0(x_1,x_0)} n(x,y)ds = \left[\int_{\Gamma^1(x_1,x_0)} n_0(x,y)ds - \int_{\Gamma^0(x_1,x_0)} n_0(x,y)ds \right] +$$
$$+ \left[\int_{\Gamma^1(x_1,x_0)} n_1(x,y)ds - \int_{\Gamma^0(x_1,x_0)} n_1(x,y)ds \right] .$$

The extremum principle clearly implies the following inequalities:

$$(9) \qquad \tau(x_1,x_0) - \int_{\Gamma^0(x_1,x_0)} n(x,y)ds \le 0 \ ,$$
$$\int_{\Gamma^1(x_1,x_0)} n_0(x,y)ds - \int_{\Gamma^0(x_1,x_0)} n_0(x,y)ds \ge 0 \ . \Bigg\}$$

Therefore in order that relation (8) hold, it is necessary that the expressions in brackets in (8) have the same order of smallness.

We determine $\tau_1(x_1,x_0)$ from (8) in the form

(9a)
$$\tau_1(x_1,x_0) = \int_{\Gamma^0(x_1,x_0)} n_1(x,y)ds + \left[\int_{\Gamma^1(x_1,x_0)} n_0(x,y)ds - \int_{\Gamma^0(x_1,x_0)} n_0(x,y)ds \right]$$
$$+ \left[\int_{\Gamma^1(x_1,x_0)} n_1(x,y)ds - \int_{\Gamma^0(x_1,x_0)} n_1(x,y)ds \right] .$$

Noting that for smooth $n_1(x,y)$ (for instance, twice continuously differentiable), the expression

(9b)
$$\int_{\Gamma^1(x_1,x_0)} n_1(x,y)ds - \int_{\Gamma^0(x_1,x_0)} n_1(x,y)ds$$

is of second order as compared to the function

(9c)
$$\int_{\Gamma^0(x_1,x_0)} n_1(x,y)ds$$

and using the result obtained above, we arrive at formula (7) on neglecting small terms of higher order. Formula (7) may also be derived through linearization of the eiconal equation

(9d)
$$|grad_{x_1} \tau(x_1,x_0)| = n(x) .$$

If we substitute the expression for n and $\tau(x_1,x_0)$ from (4) and (5) in this equation and neglect the terms n_1^2 and $|grad_{x_1}\tau_1(x_1,x_0)|^2$, we obtain

(9e)
$$(grad\ \tau_0,\ grad\ \tau_1) = n_0\ n_1$$

or in other words,

(9f)
$$\frac{d\tau_1}{ds} = n_1 .$$

This is precisely equivalent to formula (7).

Thus formula (7) holds in linear approximation. Since the function $n_0(x,y)$ is known, by solving a geometric optics problem, we can construct the two-parameter family of curves $\Gamma^0(x_1,x_0)$. The determination of $n(x,y)$ is then reduced to the construction of function $n_1(x,y)$

from $\tau_1(x_1, x_o)$. In other words, we are led to an integral-geometric
problem. In the special case where $n_o(x,y)$ is given by

$$(9g) \qquad\qquad n_o(x,y) = (ay + b)^{-1}$$

where $a > 0$ and $b \geq 0$, $\Gamma^o(x_1, x_o)$ is a circular arc with center at
the point $((x_1+x_o)/2, -b/a)$ (see [32]). The function $n_1(x,y)$ may
always be considered known for $\dot{y} = 0$ (if we pass to the limit in (7)
letting $x_1 \to x_o$, we can determine $n_1(x_o, 0)$). If we extend it in a
continuous fashion into the strip $-\frac{b}{a} \leq y \leq 0$ and then evenly about the
line $y = -\frac{b}{a}$, we wind up with a problem of determining function
$n_1(x,y)$ from its mean values over circles of arbitrary radius with
centers on the line $y = -\frac{b}{a}$, which was considered in Chapt. 1.

Another less trivial example of the use of the linearized version of
the inverse kinematic problem will be considered in the next section.

2. Application of the Linearized Version of the Inverse Kinematic Problem to Geophysics

The problem treated above may be used to give a more precise character-
ization of the global velocity distribution of longitudinal and transverse waves propagating in the earth. Under the assumption that the
earth is an elastic sphere for which the velocity of a disturbance is a
function of radius only, the data accumulated from a fairly large number
of earthquakes has been used to construct velocity distribution curves
for (longitudinal and transverse) seismic waves propagating along the
earth's radius (see, for example, [5] and [14]). A theoretical basis
for the problem and a method for solving it were given in the previously mentioned articles [14] and [37]. In principle, to solve the problem,
one should according to the method know the travel-times to any point
on the earth's surface for the seismic waves generated by an earthquake.
Actually, earthquake tremors are recorded at a network of seismological
stations situated fairly distant from one another. Therefore, to construct more reliable velocity propagation curves, one ought to average
these curves over data obtained from many earthquakes. Such work has
been done by many geophysicists and there now exists a whole collection
of velocity propagation curves for longitudinal and transverse waves.
The variations in these curves range from 10 to 15%. This is explained
by the fact that the various authors have in general made use of

different earthquakes recorded at different stations to obtain their
data. The velocity distribution curves for longitudinal and transverse
waves most widely accepted today correspond to the JEFFREYS-BULLEN
hodograph.

Meanwhile, geophysicists now have at their disposal rather reliable
information that the structure of the earth is inhomogeneous with
respect to geographic coordinates. And so the propagation velocities
of disturbances should therefore also depend on these coordinates. This
is confirmed by the fact that systematic deviations from the average
hodographs have been observed for the travel-times of waves. But the
deviations from the JEFFREYS-BULLEN hodograph, for example, are small.
Namely, for travel-times of the order of 15-20 minutes, the deviations
do not exceed 5 - 6 seconds. This provides a basis for assuming that
the propagation velocities differ little from those corresponding to
the distribution depending just on the radius.

Let $n(r,\theta,\varphi)$ be the reciprocal of the (longitudinal or transverse)
wave propagation speed. We can thus represent it in the form

$$(A) \qquad n(r,\theta,\varphi) = n_0(r) + n_1(r,\theta,\varphi) \; ,$$

where $n_0(r)$ is a known function, $n_1(r,\theta,\varphi)$ is a small function and
r,θ,φ are spherical coordinates. Let $\tau(\theta_0,\varphi_0,\theta,\varphi)$ denote the travel-
time of a seismic wave from point M_0 on the earth's surface with co-
ordinates (θ_0,φ_0) to another point M on the earth's surface with
coordinates (θ,φ) (in this connection, we are considering the earth
to be a unit sphere). We can represent it in the form

$$(B) \qquad \tau(\theta_0,\varphi_0,\theta,\varphi) = \tau_0(\theta,\varphi,\theta_0,\varphi_0) + \tau_1(\theta,\varphi,\theta_0,\varphi_0) \; ,$$

where $\tau_0(\theta_0,\varphi_0,\theta,\varphi)$ designates the travel-time from point M_0 to M
subject to a velocity distribution corresponding to the function $n(r)$,
and $\tau_1(\theta_0,\varphi_0,\theta,\varphi)$ in accordance with the above discussion is a small
function.

Observe that the curve $\Gamma^0(M,M_0)$ along which the disturbance produced
at M_0 arrives at M in the shortest time under $n = n_0(r)$ lies in
the plane of the great circle passing through M and M_0.

If we carry over the results of Sec. 1 to the three-dimensional case, we can obtain for $\tau_1(\Theta_o,\varphi_o,\Theta,\varphi)$ the following approximate formula :

(c) $$\tau_1(\Theta_o,\varphi_o,\Theta,\varphi) = \int\limits_{\Gamma^o(M,M_o)} n_1(r,\Theta,\varphi)\,ds \quad .$$

We now let the points M and M_o vary over the circumference of the unit circle. As a result, we arrive at the problem of integral geometry considered in Sec. 3 of Chapt. 1. Indeed, the curves $\Gamma^o(M,M_o)$ satisfy items 1° and 2° of the requirements imposed on the family of curves. Condition 3° will also hold if $n_o(r)$ is a twice continuously differentiable monotone function of r. The requirement that it be monotone follows from the one-dimensional velocity distribution curves, at least in the earth's mantle. The differentiability requirement can be satisfied by making small changes in $n_o(r)$ in the norm of the space C. Thus solving the integral-geometric problem, we determine $n_1(r,\Theta,\varphi)$ in each great circular cross-section of the earth. Since these circles form a two-parameter family, we have an overdeterminancy available in our problem. The function $n_1(r,\Theta,\varphi)$ can be recovered if we merely know the travel-times of waves, for example, in cross-sections passing through the earth's polar axis. The forementioned overdeterminancy can be used to average the results obtained.

In making practical use of the above procedure one must not take very many terms in the FOURIER expansion of $n_1(r,\Theta,\varphi)$ because seismological stations are situated fairly sparsely and also the resulting data is distorted by observational errors. In so doing, one will be able to allow for inhomogeneities in the earth of sufficiently large magnitude.

On the basis of our remark in Sec. 3 of Chapt. 1, we can also point out that the method may be used to clarify the local structure of the earth for those regions where there are a sufficiently large number of seismological stations and earthquakes.

CHAPTER 4

Inverse Heat Conduction Problems with Continuously Active Sources

This chapter will deal with two inverse problems for the heat equation

(1) $$a^2\frac{\partial u}{\partial t} = \Delta u + f, \qquad a = \text{const},$$

in a half-plane, where f is a function of the form

(1a) $$f(x_1,x_2,\ldots,x_n,t) = \varphi(t)f_1(x_1,x_2,\ldots,x_n) \,,$$

$\varphi(t)$ being a known function. They are then extended to the case of n-dimensional space.

1. Inverse Heat Conduction Problems for a Half-Plane

$1^{\underline{o}}$- First inverse problem: It is required to determine the function $f(x,y)$ from the equation

(2) $$a^2\frac{\partial u}{\partial t} = \Delta u + \varphi(t)f(x,y)$$

in the half-plane $y \geq 0$ under the following conditions :

(3)
$$u(x,y,0) = 0,$$
$$u(x,0,t) = h(x,t)$$
$$u(x,y_1,t) = r(x,t) \quad .$$

Let

(3a) $$v(x,y,\lambda) = \int_0^\infty e^{-\lambda^2 t} u(x,y,t)dt \,.$$

By virtue of (2) and (3), the function $v(x,y,\lambda)$ satisfies the differential equation

(4)
$$\Delta v - a^2\lambda^2 v = -\psi(\lambda)f(x,y) \,,$$
$$\psi(\lambda) = \int_0^\infty e^{-\lambda^2 t} \varphi(t)dt \quad .$$

The last two relations in (3) go over into the following :

(4a)
$$v(x,y_1,\lambda) = \int_0^\infty e^{-\lambda^2 t} r(x,t)dt = r_1(x,\lambda) \ ,$$

$$v(x,0,\lambda) = \int_0^\infty e^{-\lambda^2 t} h(x,t)dt = h_1(x,\lambda) \ .$$

A fundamental solution for equation (4) is the HANKEL function

(4b) $\frac{1}{2\pi} H_0^{(1)} (ia\lambda r) = \frac{1}{2\pi} K_0(a\lambda r)$, $r = \sqrt{x^2+y^2}$.

It satisfies this equation everywhere except at $r = 0$.

The GREEN'S function of first kind is expressed in terms of the fundamental solution by

(4c) $G(x,y;\xi,\eta) = \frac{1}{2\pi} [K_0(a\lambda R_1) - K_0(a\lambda R_2)|$,

$R_1 = [(x-\xi)^2+(y-\eta)^2]^{\frac{1}{2}}$, $R_2 = [(x-\xi)^2+(y+\eta)^2]^{\frac{1}{2}}$.

Therefore the solution to (4) may be represented in the form

(5) $v(x,y\lambda) = \frac{a\lambda y}{\pi} \int_{-\infty}^\infty \frac{1}{R_0} K_1(a\lambda R_0) \ h_1(\xi,\lambda)d\xi$

$+\psi(\lambda) \int_0^\infty \int_{-\infty}^\infty G(x,y;\xi,\eta) \ f(\xi,\eta)d\xi d\eta$,

$R_0 = [(x-\xi)^2 + y^2]^{\frac{1}{2}}$.

Setting $y = y_1$ in this last equation, we obtain an integral equation of first kind for the unknown function $f(\xi,\eta)$, namely

$$\int_0^\infty \int_{-\infty}^\infty [K_0(a\lambda R_3) - K_0(a\lambda R_4)] f(\xi,\eta)d\xi d\eta = g(x,\lambda),$$

$R_3 = [(x-\xi)^2+(\eta-y_1)^2]^{\frac{1}{2}}$; $R_4 = [(x-\xi)^2+(\eta+y_1)^2]^{\frac{1}{2}}$;

(6)

$g(x,\lambda) = \frac{2}{\psi(\lambda)} [\pi r_1(x,\lambda) - a\lambda y_1 \int_{-\infty}^\infty \frac{1}{R} K_1(a\lambda R)h_1(\xi,\lambda)d\xi]$,

$R = [(x-\xi)^2+y_1^2]^{\frac{1}{2}}$.

The right-hand side $g(x,\lambda)$ of equation (6) is subject to the sole condition that the equation have as solution a function $f(x,y)$ such that:

1. $f(x,y) = 0$ for $y < y_1$, $y_1 > 0$;
2. $f(x,y) \in L_1(D)$, $D = \{-\infty < x < \infty; \; 0 < y < \infty\}$.

Taking FOURIER transforms in equation (6) with respect to x and using the relation (see [7])

$$(7) \qquad \int_0^\infty K_0[b(c^2+t^2)^{\frac{1}{2}}] \cos ut \; dt = \frac{\pi}{2\sqrt{b^2+u^2}} \; e^{-|c|\sqrt{b^2+u^2}}$$

we obtain

$$(7a) \qquad \int_0^\infty F(\omega,n)\left\{e^{-|n-y_1|\sqrt{a^2\lambda^2+\omega^2}} - e^{-|n+y_1|\sqrt{a^2\lambda^2+\omega^2}}\right\}dn$$

$$= \frac{\sqrt{a^2\lambda^2+\omega^2}}{\pi\sqrt{2\pi}} \int_{-\infty}^\infty g(x,\lambda)\, e^{-i\omega x} \; dx \; ,$$

with

$$(7b) \qquad F(\omega,n) = \frac{1}{\sqrt{2\pi}} \int_{-\infty}^\infty f(\xi,n)\, e^{-i\omega\xi} \; d\xi \; .$$

Taking into consideration that $f(x,y) = 0$ for $y < y_1$, we can write this last equation in the form

$$\int_{y_1}^\infty F(\omega,n)\, e^{-n\sqrt{a^2\lambda^2+\omega^2}} \; dn = Q(\omega,\lambda) \; ,$$

$$(8) \qquad Q(\omega,\lambda) = \frac{\sqrt{a^2\lambda^2+\omega^2}}{2\pi \sinh y_1 \sqrt{a^2\lambda^2+\omega^2}} \; G(\omega,\lambda) \; ,$$

$$G(\omega,\lambda) = \frac{1}{\sqrt{2\pi}} \int_{-\infty}^\infty g(x,\lambda)\, e^{-i\omega x} \; dx \; .$$

Let us show that we have the right to take FOURIER transforms in equation (6) with respect to x under the above assumptions on $f(x,y)$.

From (7) it follows that the kernel of the integral equation (6) is an absolutely integrable function. Then by the theorem convolution for two absolutely integrable functions, we infer that $g(x,\lambda)$ is also absolutely integrable in the argument x and hence $G(\omega,\lambda)$ exists. It is not hard to derive for $G(\omega,\lambda)$ the estimate

$$(9) \qquad |G(\omega,\lambda)| < \frac{2\pi \sinh y_1 \sqrt{a^2\lambda^2+\omega^2}}{\sqrt{a^2\lambda^2+\omega^2}} \; e^{-y_1\sqrt{a^2\lambda^2+\omega^2}} \int_{y_1}^\infty |F(\omega,n)|\,dn \; .$$

The fact that $f(\xi,\eta) \in L_1(D)$ implies

$$(9a) \qquad |F(\omega,\eta)| \leq \frac{1}{\sqrt{2\pi}} \int_{-\infty}^{\infty} |f(\xi,\eta)| d\xi ,$$

$$\int_{y_1}^{\infty} |F(\omega,\eta)| d\eta \leq \frac{1}{\sqrt{2\pi}} \int_{y_1}^{\infty}\int_{-\infty}^{\infty} |f(\xi,\eta)| d\xi d\eta = \alpha < \infty .$$

Substituting the last inequality into (9), we finally obtain

$$(10) \qquad |G(\omega,\lambda)| < \frac{\alpha\pi}{\sqrt{a^2\lambda^2+\omega^2}} .$$

Introduce the notation

$$t = \eta - y_1 , \qquad p^2 = a^2\lambda^2 + \omega^2 ,$$

$$(10a) \qquad F_1(\omega,t) = F(\omega,t+y_1) = F_{11}(\omega,t) + iF_{12}(\omega,t) ,$$

$$Q_1(\omega,p) = e^{y_1 p} Q(\omega, \frac{1}{a}\sqrt{p^2-\omega^2}) = Q_{11}(\omega,p) + iQ_{12}(\omega,p) .$$

Then equation (8) can be split into two independent integral equations for the unknown functions F_{11} and F_{12}, namely,

$$(11) \qquad \int_0^{\infty} F_{1l}(\omega,t) e^{-pt} dt = Q_{1l}(\omega,p) , \qquad (l=1,2) .$$

The functions F_{11} and F_{12} are clearly continuous and bounded since $f(\xi,\eta) \in L_1(D)$. Hence it follows that each of the equations in (11) has just one solution and it may be expressed by the formula [26]:

$$(12) \qquad F_{1l}(\omega,z) = \lim_{n\to\infty} \frac{(-1)^n (\frac{n}{z})^{n+1} \frac{\partial^n}{\partial p^n} Q_{1l}(\omega,\frac{n}{z})}{n!} , \qquad (l=1,2) .$$

Further, it is known that if the FOURIER transform

$$(13) \qquad F(\omega,\eta) = \frac{1}{\sqrt{2\pi}} \int_{-\infty}^{\infty} f(\xi,\eta) e^{-i\omega\xi} d\xi$$

of a summable function $f(\xi,\eta)$ (for fixed η) is equal to zero for all ω, then $f(\xi,\eta) = 0$ almost everywhere. Therefore, the unique solution to equation (13) is given by

$$(13a) \qquad f(\xi,\eta) = \frac{1}{\sqrt{2\pi}} \int_{-\infty}^{\infty} F(\omega,\eta) e^{i\omega\xi} d\omega .$$

Thus the inverse problem formulated above for equation (2) has at most one solution.

2°- <u>Second inverse problem</u>: We reduced the inverse problem (2), (3) to integral equation (8) with the help of formula (5) which was the solution to the DIRICHLET problem for equation (4). An integral equation analogous to (8) may be derived by considering the NEUMANN problem for the same equation.

Let $u(x,y,t)$ be the solution to equation (2) in the half-plane $y \geq 0$ such that

$$
\begin{aligned}
&u(x,y,0) = 0 \ , \\
(14) \qquad &u(x,0,t) = h(x,t) \ , \\
&\frac{\partial}{\partial y} u(x,0,t) = r(x,t) \ .
\end{aligned}
$$

Taking LAPLACE transforms with respect to t in equation (2) we wind up with equation (4) for v.

The boundary conditions in (14) go over into the following :

$$
\begin{aligned}
(14a) \qquad &v(x,0,\lambda) = \int_0^\infty e^{-\lambda^2 t} h(x,t)dt = \varphi_1(x,\lambda) \ , \\
&\frac{\partial}{\partial y} v(x,0,\lambda) = \int_0^\infty e^{-\lambda^2 t} r(x,t)dt = \varphi_2(x,\lambda) \ .
\end{aligned}
$$

It is easy to show that the function

$$
(14b) \qquad N(x,y;\xi,\eta) = \frac{1}{2\pi} [K_o(a\lambda R_1) + K_o(a\lambda R_2)],
$$

$$
R_1 = [(x-\xi)^2+(y-\eta)^2]^{\frac{1}{2}}; \quad R_2 = [(x-\xi)^2+(y+\eta)^2]^{\frac{1}{2}}
$$

satisfies the differential equation (4) everywhere except at $x = \xi$, $y = \eta$ where it has a logarithmic singularity. The normal derivative of this function vanishes along the boundary of the half-plane. Hence, $N(x,y;\xi,\eta)$ is the GREEN'S function of second kind for the half-plane. The solution to equation (4) can thus be represented in the following form :

$$v(x,y,\lambda) = \frac{1}{\pi} \int_{-\infty}^{\infty} K_o(a\lambda R_o)\, \varphi_2(\xi,\lambda)d\xi$$

(15)
$$+ \psi(\lambda)\int_0^{\infty}\int_{-\infty}^{\infty} N(x,y;\xi,\eta)\, f(\xi,\eta)d\xi\, d\eta \ ,$$

$$R_o = [(x-\xi)^2+y^2]^{\frac{1}{2}} \ .$$

Setting $y = 0$ in (15), we obtain the integral equation

(16)
$$\int_0^{\infty}\int_{-\infty}^{\infty} K_o\{a\lambda[(x-\xi)^2+\eta^2]^{\frac{1}{2}}\}f(\xi,\eta)d\xi d\eta = g(x,\lambda) \ ,$$

$$g(x,\lambda) = \frac{1}{\psi(\lambda)} \left[\pi\varphi_1(x,\lambda) - \int_{-\infty}^{\infty} K_o(a\lambda|x-\xi|)\, \varphi_2(\xi,\eta)d\xi\right]$$

for the unknown function $f(\xi,\eta)$. If we take FOURIER transforms with respect to x, we wind up with

(17)
$$\int_0^{\infty} e^{-\eta\sqrt{a^2\lambda^2+\omega^2}}\, F(\omega,\eta)d\eta = \frac{1}{\pi}\sqrt{a^2\lambda^2+\omega^2}\ G(\omega,\lambda)$$

$$G(\omega,\lambda) = \frac{1}{\sqrt{2\pi}}\int_{-\infty}^{\infty} g(x,\lambda)\ e^{-i\omega x}dx \ .$$

Introduce the notation

$$p^2 = a^2\lambda^2+\omega^2,$$

(17a)
$$\Phi(p,\omega) = \frac{1}{\pi}\, pG(\omega, \frac{1}{a}\sqrt{p^2-\omega^2}) = \Phi_1(p,\omega)+i\, \Phi_2(p,\omega) \ ,$$

$$F(\omega,\eta) = F_1(\omega,\eta)+iF_2(\omega,\eta) \ .$$

Separating real and imaginary parts in (17), we obtain two independent integral equations for the unknown functions F_1 and F_2 :

(18)
$$\int_0^{\infty} e^{-ph}F_1(\omega,\eta)d\eta = \Phi_1(\omega,p) \ .$$

To justify taking FOURIER transforms with respect to x in (16), we may subject $g(x,\lambda)$ to the condition that the solution to (16) be a certain function $f(x,y)\in L_1(D)$, $D = \{-\infty < x < \infty; 0 < y < \infty\}$. We can then represent the unique solution to equation (17) by means of formula (12). On inverting the FOURIER transform (13), we arrive at the unique solution $f(x,y)$ to the inverse problem (2), (14).

We point out that the solution to the integral equation (18) may be expressed differently. If we suppose that

1. $f(x,y) \in L_1 (-\infty < x < \infty)$, $y \in (0,\infty)$,
2. $F(\omega,z) \in L_2 (0 < z < \infty)$, $\omega \in (-\infty,\infty)$,

then the solution to (18) has the following form [26]:

$$(18a) \quad F_1(\omega,z) = \underset{A \to \infty}{\text{l.i.m.}} \; \frac{1}{2\pi z} \int_0^\infty \phi_1\left(\omega,\frac{t}{z}\right) \frac{dt}{\sqrt{t}} \int_{-A}^A \frac{t^1 \, \varsigma d\xi}{\Gamma(1\xi + \frac{1}{2})}$$

where $\Gamma(x)$ is the gamma-function.

2. n-Dimensional Inverse Heat Conduction Problems

We shall consider the same problems now as in the preceding section but for n-dimensional space. Since the reasoning does not involve any essential changes, our presentation will be as condensed as possible.

1^0- First inverse problem: It is required to find the function $f(x,x_n)$, $x = (x_1,x_2,\ldots, x_{n-1})$, from the equation

$$(1) \qquad\qquad a^2 \frac{\partial u}{\partial t} = \Delta u + \varphi(t) f(x,x_n)$$

in the n-dimensional half-space $D = \{x_n \geq 0\}$ providing the following are given for $u(x,x_n,t)$:

$$(2) \qquad\qquad u(x,x_n,0) = 0,$$
$$u(x,0,t) = h(x,t),$$
$$u(x,a,t) = r(x,t) , \quad a = \text{const.}$$

Direct verification shows that the function

$$(3) \qquad\qquad v(x,x_n,\lambda) = \int_0^\infty e^{-\lambda^2 t} u(x,x_n,t) dt$$

satisfies the differential equation

$$(4) \qquad\qquad \Delta v - a^2 \lambda^2 v = -\psi(\lambda) f(x,x_n), \quad \Delta = \frac{\partial^2}{\partial x_1^2} + \ldots + \frac{\partial^2}{\partial x_n^2} ,$$

$$\psi(\lambda) = \int_0^\infty e^{-\lambda^2 t} \varphi(t) dt$$

and relations

$$(5) \qquad v(x,0,\lambda) = \int_0^\infty e^{-\lambda^2 t} h(x,t)dt = h_1(x,\lambda) \ ,$$

$$v(x,\alpha,\lambda) = \int_0^\infty e^{-\lambda^2 t} r(x,t)dt = r_1(x,\lambda) \ \ .$$

Let $K_{\frac{n-2}{2}}(x)$ be the cylindrical HANKEL function of imaginary argument.
The function

$$(6) \qquad Q_\lambda(x,x_n;\xi,\xi_n) = \frac{1}{2\pi} \left(\frac{a\lambda}{2\pi R}\right)^{\frac{n-2}{2}} K_{\frac{n-2}{2}} (a\lambda R) \ ,$$

$$R = \left[(x_1-\xi_1)^2 + \ldots + (x_n-\xi_n)^2\right]^{\frac{1}{2}}, \cdot \qquad \xi = (\xi_1,\xi_2,\ldots,\xi_{n-1})$$

with singularity at (ξ,ξ_n) is a fundamental solution of equation (4).

The GREEN'S function of first kind or fundamental solution for (4),
being a function vanishing along the boundary of the half-space, can be
expressed by the formula

$$(6a) \quad G_\lambda(x,x_n;\xi,\xi_n) = Q_\lambda(x,x_n;\xi,\xi_n) - Q_\lambda(x,x_n;\xi,-\xi_n) \ \ .$$

The solution to equation (4) is then expressible as

$$v(x,x_n,\lambda) = -\int_{S_n} h_1(\xi,\lambda) \frac{\partial}{\partial \xi_n} Q_\lambda(x,x_n;\xi,0)d\xi$$

$$(7)$$

$$+ \ \psi(\lambda)\int_D f(\xi,\xi_n) \ G_\lambda(x,x_n;\xi,\xi_n)d\xi d\xi_n \ \ ,$$

where S_n is the surface of the half-space D.

Setting $x_n = \alpha$ in relation (7), we arrive at the integral equation

$$\int_D f(\xi,\xi_n) \ G_\lambda(x,\alpha;\xi,\xi_n)d\xi d\xi_n = g(x,\lambda) \ ,$$

$$(8)$$

$$g(x,\lambda) = \frac{1}{\psi(\lambda)} \left[r_1(x,\lambda) + \int_{S_n} h_1(\xi,\lambda) \frac{\partial}{\partial \xi_n} Q_\lambda(x,\alpha;\xi,0)d\xi\right]$$

for the unknown function $f(x,x_n)$.

Suppose that for given $g(x,\lambda)$, the solution to equation (8) is a function $f(x,x_n)$ satisfying the conditions

1. $f(x,x_n) = 0$ for $x_n < \alpha$, $\alpha > 0$,
2. $f(x,x_n) \in L_1(D)$.

Let us show that equation (8) can have at most one solution under these conditions.

We first evaluate the integral

$$(8a) \qquad J_n = (\frac{1}{\sqrt{2\pi}})^{n-1} \int_{-\infty}^{\infty}\dots\int_{-\infty}^{\infty} Q_\lambda(x,\alpha;0,0)\; e^{-i\sum_{k=1}^{n-1}\omega_k x_k}dx_1,dx_2\dots dx_{n-1}.$$

Since Q_λ is even in all variables x_1, the integral may be written as

$$J_n = 2^{n-1}(2\pi)^{\frac{1-2n}{2}}(a\lambda)^{\frac{n-2}{2}}\int_0^{\infty}\dots\int_0^{\infty} R^{-\frac{n-2}{2}} K_{\frac{n-2}{2}}(a\lambda R)$$

$$(9)$$

$$\times \prod_{k=1}^{n-1}\cos\omega_k x_k dx_1 dx_2\dots dx_{n-1}\;;\; R = [x_1^2 + x_2^2 +\dots+ x_{n-1}^2 + a^2]^{\frac{1}{2}}.$$

Applying the formula [7]

$$\int_0^{\infty} (\sqrt{t^2+c^2})^{-\frac{n+2}{2}} K_{\frac{n-2}{2}}[b\sqrt{c^2+t^2}]\cos ut\, dt$$

$$=(\frac{\pi}{2b})^{\frac{1}{2}} (b|c|)^{\frac{3-n}{2}} (\sqrt{b^2+u^2})^{\frac{n-3}{2}} K_{\frac{n-3}{2}}[|c|\sqrt{b^2+u^2}]$$

to (9), we obtain the following representation:

$$(10) \qquad J_n = (2\pi)^{\frac{1-n}{2}}\; \frac{1}{2\sqrt{a^2\lambda^2+\omega_1^2+\dots+\omega_{n-1}^2}}\; e^{-|\alpha|\sqrt{a^2\lambda^2+\omega_1^2+\dots+\omega_{n-1}^2}}.$$

The integral equation (8) is obviously of convolution-type in the variables $x_1,x_2,\dots,x_{n-1}$. From (10) it follows that the kernel of the equation is an absolutely integrable function. Since $f(x,x_n)$ is likewise absolutely integrable in domain D, applying to equation (8) the convolution theorem for two absolutely integrable functions and making use of (10), we obtain

$$(11) \qquad \int_0^{\infty} F(\xi_n,\omega)\left\{ e^{-|\xi_n-\alpha|\sqrt{a^2\lambda^2+|\omega|^2}} - e^{-|\xi_n+\alpha|\sqrt{a^2\lambda^2+|\omega|^2}}\right\}d\xi_n$$

$$= 2^{n-1}\sqrt{a^2\lambda^2+|\omega|^2}\; G(\omega,\lambda)\;,$$

where

$$\omega \overset{!}{=} (\omega_1, \omega_2, \ldots, \omega_{n-1}), \qquad |\omega|^2 = \omega_1^2 + \omega_2^2 + \ldots + \omega_{n-1}^2 ,$$

$$(11a) \quad G(\omega, \lambda) = \left(\frac{1}{\sqrt{2\pi}} \right)^{n-1} \int_{-\infty}^{\infty} \ldots \int_{-\infty}^{\infty} g(x, \lambda) \, e^{-i \sum_{k=1}^{n-1} \omega_k x_k} \, dx ,$$

$$F(x_n, \omega) = \left(\frac{1}{\sqrt{2\pi}} \right)^{n-1} \int_{-\infty}^{\infty} \ldots \int_{-\infty}^{\infty} f(x, x_n) \, e^{-i \sum_{k=1}^{n-1} \omega_k x_k} \, dx .$$

By the first assumption, $f(x, x_n)$ vanishes for $x_n < \alpha$ and so $F(\xi_n, \omega)$ vanishes for $\xi_n < \alpha$. Therefore equation (11) has the simpler form

$$(12) \quad \int_{\alpha}^{\infty} F(\xi_n, \omega) \, e^{-\xi_n \sqrt{a^2 \lambda^2 + |\omega|^2}} \, d\xi_n = \frac{2^{n-2} \sqrt{a^2 \lambda^2 + |\omega|^2}}{\sinh(\alpha \sqrt{a^2 \lambda^2 + |\omega|^2})} \, G(\omega, \lambda) .$$

We have obtained exactly the same integral equation as in the two-dimensional case. Its solution may be found using formula (12) of Sec.1 of this chapter.

Repeating the reasoning of Sec. 1, we arrive at the conclusion that the solution to the inverse problem (1), (2) is unique in the class of summable functions.

2^{0}- Second inverse problem: Consider the inverse problem (1) and (2) but with the condition

$$(12a) \qquad\qquad u(x, \alpha, t) = r(x, t)$$

replaced by

$$(12b) \qquad\qquad \frac{\partial}{\partial x_n} u(x, 0, t) = m(x, t) .$$

Eliminating the variable t from equation (1) by application of the transformation (3), we obtain for $v(x, x_n, \lambda)$ equation (4) and the relations

$$(12c) \quad \begin{aligned} v(x, 0, \lambda) &= \int_0^{\infty} e^{-\lambda^2 t} h(x, t) \, dt = h_1(x, \lambda) \\[2mm] \frac{\partial}{\partial x_n} v(x, 0, \lambda) &= \int_0^{\infty} e^{-\lambda^2 t} m(x, t) \, dt = m_1(x, \lambda) . \end{aligned}$$

Direct verification shows that the expression

(12d) $\quad N_\lambda(x,x_n;\xi,\xi_n) = Q_\lambda(x,x_n;\xi,\xi_n) + Q_\lambda(x,x_n;\xi,-\xi_n)$

is the GREEN'S function of second kind for the half-space D. By using it we can represent the solution to (4) by

(13)
$$v(x,x_n,\lambda) = -2 \int_{S_n} m_1(x,\lambda)\ Q_\lambda(x,x_n;\xi,0)d\xi$$
$$+ \psi(\lambda) \int_D f(\xi,\xi_n)\ N_\lambda(x,x_n;\xi,\xi_n)d\xi \quad .$$

If we set $x_n = 0$ in this last expression, we wind up with an integral equation for $f(\xi,\xi_n)$, namely

(14)
$$\int_D f(\xi,\xi_n)\ Q_\lambda(x,0;\xi,\xi_n)d\xi d\xi_n = g(x,\lambda),$$
$$g(x,\lambda) = \frac{1}{\psi(\lambda)}\left\{\frac{1}{2}h_1(x,\lambda) + \int_{S_n} m_1(x,\lambda)\ Q_\lambda(x,0;\xi,0)d\xi\right\}.$$

Suppose the right-hand side $g(x,\lambda)$ of equation (14) is such that the equation has as solution a function $f(x,x_n) \in L_1(D)$.

Taking FOURIER transforms in (14) with respect to $x_1,x_2,\dots x_{n-1}$ and making use of (10), we obtain the integral equation

(14a) $\quad \displaystyle\int_0^\infty F(\xi_n,\omega)\ e^{-\xi_n\sqrt{a^2\lambda^2+|\omega|^2}}d\xi_n = 2^{n-1}\sqrt{a^2\lambda^2+|\omega|^2}\,G(\omega,\lambda)\quad .$

Hence the inverse problem with CAUCHY data on the boundary of the half-space **D** can have at most one solution.

3. Application of the Problems to Geophysics

As we know, there is a large range of mathematical physics problems dealing with heating or cooling of bodies containing internal sources of heat.

We point out, for example, the problem of the effect of radioactive decay on the temperature of the earth's crust [36]. The gist of this problem is as follows.

Radioactive decay of elements causes the earth's crust to heat up, its temperature satisfying the heat equation

(14b)

$$a^2 \frac{\partial u}{\partial t} = \Delta u + f ,$$

$$f = \varphi(t) f_1(x,y,z) .$$

The function f_1 characterizes the volumetric thermal source strength, and $\varphi(t)$ is given by

(14c)
$$\varphi(t) = \alpha e^{-\lambda t}$$

where λ is the half-life of the corresponding radioactive element.

Thus knowing, for example, the functions

(14d)
$$u(x,y,0,t) = n(x,y,t) ,$$
$$\frac{\partial}{\partial z} u(x,y,0,t) = m(x,y,t) ,$$

we can determine the volumetric strength of radioactive elements scattered in the earth's crust under the conditions specified above.

CHAPTER 5

Inverse Problems for Second-Order Elliptic Equations

Let v be a function satisfying the differential equation

(1) $\Delta v = (a+\lambda b)v,\ a(P) \geq 0,\ a(P)+\lambda b(P) \geq 0,$

$$P = (x_1, x_2, \ldots, x_n)$$

in a domain D under certain boundary conditions. Here a and b are
bounded continuous functions and λ is a parameter.

The following three types of boundary conditions are usually considered:

1) v takes on prescribed values on the boundary S of D :

(1a) $v\big|_S = f$

2) the normal derivative of v is prescribed on S :

(1b) $\frac{\partial v}{\partial n}\big|_S = \varphi$

3) v satisfies on S the condition

(1c) $[\frac{\partial v}{\partial n} + hv]_S = \psi,$ $h = \text{const} > 0$

where h and the function ψ are prescribed.

Problems 1) – 3) may be solved with the help of the GREEN'S functions
for equation (1) in D. The GREEN'S function G(P,Q) of first kind is
defined as the fundamental solution of equation (1) vanishing on the
boundary S of domain D. The GREEN'S function N(P,Q) of second kind
is the fundamental solution of (1) whose normal derivative vanishes on
S. Finally, the GREEN'S function of third kind is the fundamental
solution R_h of the equation for which

(1d) $\left[\frac{\partial R_h}{\partial n} + hR_h\right]_S = 0$.

Apart from the direct problems for (1) involving the determination of
a solution under one of the particular boundary conditions, of interest

in a certain sense are the inverse problems dealing with the determination of the function $b(P)$ from certain properties of the solutions
to the equation. In this connection, one can set up various inverse
problems depending on the nature of the information known about the
solutions to equation (1). In this chapter, we shall stop to consider
one such formulation.

Let $G_1(P,Q)$ and $G_2(P,Q')$ be the GREEN'S functions of first kind for
equation (1) in D corresponding to $\lambda=\lambda_1$ and $\lambda=\lambda_2$. We cut out of D
two infinitely small spheres described around the points Q and Q'.
Denote the resultant domain by D_1. Applying GREENS'S theorem to $G_1(P,Q)$
and $G_2(P,Q')$ in D_1, we have

$$(2) \qquad G_1(Q,Q') - G_2(Q,Q') = (\lambda_2-\lambda_1)\int_D b(P)G_1(P,Q)G_2(P,Q')dP \ .$$

Analogous relations also hold for the other GREEN'S functions. In particular, when $\lambda_2=\lambda$ and $\lambda_1=0$ we obtain a FREDHOLM equation of second
kind for G :

$$(3) \qquad G(Q,Q') - G_\lambda(Q,Q') = \lambda\int_D b(P)G(P,Q)G_\lambda(P,Q')dP \ .$$

For sufficiently small λ its solution is an analytic function of λ.
Therefore differentiating (3) with respect to λ and setting $\lambda=0$, we
have

$$(4) \qquad \frac{\partial G_\lambda(Q,Q')}{\partial\lambda}\bigg|_{\lambda=0} = -\int_D b(P)G(P,Q)G(P,Q')dP \ .$$

This may be regarded as an integral equation of first kind for the
function $b(P)$.

Below we shall consider some specific inverse problems for equation (1).

1. Inverse Problem for Equation (1) in a Half-Plane

Let $a(P) = a^2 = $ const., $P = (\xi,\eta)$, $b(P) = 0$ for $\eta < y_1$, and let the
domain D be the half-plane $\eta \geq 0$. Under these conditions, equation
(4) becomes

$$(5) \qquad \int_0^\infty \int_{-\infty}^\infty b(\xi,\eta) [K_0(ar_1)-K_0(ar_2)] [K_0(ar_3)-K_0(ar_4)]d\xi d\eta = f(x_1,x_2)$$

where $K_0(ar)$ is the HANKEL function of imaginary argument and

(5a)
$$r_1 = [(\xi-x_1)^2+(\eta-y_1)^2]^{\frac{1}{2}}, \quad r_2 = [(\xi-x_1)^2+(\eta+y_1)^2]^{\frac{1}{2}},$$
$$r_3 = [(\xi-x_2)^2+(\eta-y_1)^2]^{\frac{1}{2}}, \quad r_4 = [(\xi-x_2)^2+(\eta+y_1)^2]^{\frac{1}{2}},$$
$$y_1 = \text{const} > 0 .$$

We impose on the right-hand side $f(x_1,x_2)$ of integral equation (5) the single requirement that the solution $b(\xi,\eta)$ of the equation belong to $L_1(D)$.

We take FOURIER transforms in (5) with respect to x_1 and x_2 using in this connection equation (7) of Sec.1, Chapt.4.

Equation (5) then assumes the form

$$\int_{y_1}^{\infty} \int_{-\infty}^{\infty} b(\xi,\eta) \, e^{-i(\omega_1+\omega_2)\xi} \, e^{-\eta(\sqrt{a^2+\omega_1^2} + \sqrt{a^2+\omega_2^2})} d\xi d\eta = F_1(\omega_1,\omega_2),$$

(6)
$$F_1(\omega_1,\omega_2) = \frac{\sqrt{(a^2+\omega_1^2)(a^2+\omega_2^2)}}{2\pi \sinh y_1\sqrt{a^2+\omega_1^2} \, \sinh y_1\sqrt{a^2+\omega_2^2}} F(\omega_1,\omega_2) ,$$

$$F(\omega_1,\omega_2) = \frac{1}{2\pi} \int_{-\infty}^{\infty} \int_{-\infty}^{\infty} f(x_1,x_2) \, e^{-i(\omega_1 x_1+\omega_2 x_2)} dx_1 dx_2 .$$

It is not hard to show that

(6a)
$$|F(\omega_1,\omega_2)| < \frac{\pi}{2\sqrt{(a^2+\omega_1^2)(a^2+\omega_2^2)}} \int_{y_1}^{\infty} \int_{-\infty}^{\infty} |b(\xi,\eta)| d\xi d\eta$$

which implies that it is possible to take FOURIER transforms in (5).

In what follows we shall regard ω_1 positive and ω_2 negative. Let

(7)
$$u = \omega_1+\omega_2 , \quad u \in (-\infty,\infty)$$
$$v = \sqrt{a^2+\omega_1^2} + \sqrt{a^2+\omega_2^2} , \quad v \in (2a,\infty) .$$

The JACOBIAN

(7a)
$$\frac{\partial(u,v)}{\partial(\omega_1,\omega_2)} = \frac{\omega_2}{\sqrt{a^2+\omega_2^2}} - \frac{\omega_1}{\sqrt{a^2+\omega_1^2}} \neq 0$$

is continuous for all pertinent ω_1 and ω_2. Hence, the mapping $(\omega_1,\omega_2) \to (u,v)$ is one-to-one and so has an inverse.

Substituting (7) in (6), we obtain

$$(8) \qquad \int_{y_1}^{\infty} e^{-\eta v} r(u,\eta) d\eta = F_2(u,v) ,$$

$$(9) \qquad r(u,\eta) = \int_{-\infty}^{\infty} b(\xi,\eta) e^{-iu\xi} d\xi .$$

Since $b(\xi,\eta)$ is by hypothesis an absolutely integrable function, $r(u,\eta)$ is continuous and bounded. Therefore, the unique solution to (8) may be expressed in the form [26]

$$(9a) \qquad r(u, y_1+t) = \lim_{n\to\infty} \frac{(-1)^n (\frac{n}{t})^{n+1} \frac{\partial^n}{\partial v^n} F_3(u,\frac{n}{t})}{n!} ,$$

$$F_3(u,v) = e^{y_1 v} F_2(u,v) .$$

It is known that equation (9) has a unique solution (almost everywhere) in the class of absolutely integrable functions. The solution may be represented by

$$(10) \qquad b(\xi,\eta) = \frac{1}{2\pi} \int_{-\infty}^{\infty} r(u,\eta) e^{iu\xi} du .$$

Thus we have proved the following uniqueness theorem for equation (1) in a half-plane.

Theorem: *The inverse problem for equation (1) has at most one solution in the class of absolutely integrable functions.*

The entire above discussion is clearly valid for the GREEN'S function of second kind, which may be expressed in the form

$$(10a) \qquad N(P,Q) = \frac{1}{2\pi} [K_0(ar_1) + K_0(ar_2)] .$$

2. Inverse Problem for Equation (1) in a Half-Space

We shall assume $a(P) = a^2 = $ const., $P = (\xi,\eta,\zeta)$, $b(P) = 0$ for $\zeta < z_1$ and the domain D to be the half-space $\zeta \geq 0$.

Equation (4) thus assumes the form

$$\int\limits_{Z_1}^{\infty}\int\limits_{-\infty}^{\infty}\int\limits_{-\infty}^{\infty} b(\xi,\eta,\varsigma)(\frac{e^{-aR_1}}{R_1} - \frac{e^{-aR_2}}{R_2})(\frac{e^{-aR_3}}{R_3} - \frac{e^{-aR_4}}{R_4})d\xi d\eta d\varsigma = f(x_1,x_2,y_1,y_2),$$

(1)
$$R_1 = [(\xi-x_1)^2+(\eta-y_1)^2+(\varsigma-z_1)^2]^{\frac{1}{2}}; \quad R_2 = [(\xi-x_1)^2+(\eta-y_1)^2+(\varsigma+z_1)^2]^{\frac{1}{2}};$$

$$R_3 = [(\xi-x_2)^2+(\eta-y_2)^2+(\varsigma-z_1)^2]^{\frac{1}{2}}; \quad R_4 = [(\xi-x_2)^2+(\eta-y_2)^2+(\varsigma+z_1)^2]^{\frac{1}{2}};$$

$$z_1 = \text{const} > 0;$$

Let the function $f(x_1,x_2,y_1,y_2)$ be such that the solution $b(\xi,\eta,\varphi)$ of equation (1) belongs to $L_1(D)$.

As before, on taking FOURIER transforms with respect to x_1,x_2,y_1 and y_2 in (1), we obtain

(2)
$$\int\limits_{Z_1}^{\infty}\int\limits_{-\infty}^{\infty}\int\limits_{-\infty}^{\infty} b(\xi,\eta,\varsigma)e^{-1(\omega_1+\omega_2)\xi - 1(\omega_2+\omega_4)\eta}$$

$$\cdot e^{-\varsigma(\sqrt{a^2+\omega_1^2+\omega_2^2} + \sqrt{a^2+\omega_3^2+\omega_4^2})}d\xi d\eta d\varsigma$$

$$= \frac{\sqrt{a^2+\omega_1^2+\omega_2^2}\,\sqrt{a^2+\omega_3^2+\omega_4^2}\,F(\omega_1,\omega_2,\omega_3,\omega_4)}{4\sinh z_1\sqrt{a^2+\omega_1^2+\omega_2^2}\,\sinh z_1\sqrt{a^2+\omega_3^2+\omega_4^2}}, \quad F(\omega_1,\omega_2,\omega_3,\omega_4)$$

$$= \frac{1}{(2\pi)^2}\int\limits_{-\infty}^{\infty}\int\limits_{-\infty}^{\infty} f(x_1,x_2,y_1,y_2)e^{-1(\omega_1 x_1+\omega_2 y_1+\omega_3 x_2+\omega_4 y_2)}dx_1 dx_2 dy_1 dy_2 .$$

It is not hard to show that the FOURIER transform of $f(x_1,x_2,y_1,y_2)$ exists and therefore one may consider equation (2) instead of (1).

Let $\omega_1 > 0$, $\omega_3 < 0$, $\omega_4 = 0$ and $\omega_2 \in (-\infty,\infty)$ and introduce the notation

(3)
$$u = \omega_1 + \omega_3, \quad u \in (-\infty,\infty),$$
$$v = \omega_2, \quad v \in (-\infty,\infty),$$
$$w = \sqrt{a^2+\omega_1^2+\omega_2^2} + \sqrt{a^2+\omega_3^2}, \quad w \in (2a,\infty).$$

The JACOBIAN

(3a)
$$\frac{\partial(u,v,w)}{\partial(\omega_1,\omega_2,\omega_3)} = \frac{\omega_3}{\sqrt{a^2+\omega_3^2}} - \frac{\omega_1}{\sqrt{a^2+\omega_1^2+\omega_2^2}} \neq 0$$

is continuous for the considered values of ω_1 . Therefore, the mapping $(\omega_1, \omega_2, \omega_3) \to (u,v,w)$ is one-to-one and has an inverse.

Substituting (3) into (2), we have

(4)
$$\int_{z_1}^{\infty} \int_{-\infty}^{\infty} \int_{-\infty}^{\infty} b(\xi, n, \zeta) \, e^{-iu\xi - ivn} e^{-\xi w} d\xi dn d\zeta = F_2(u,v,w) \, ,$$

$$F_2(u,v,w) = F_1\left[\omega_1(u,v,w), \, \omega_2(u,v,w), \, \omega_3(u,v,w)\right] \, .$$

Let

(5)
$$\int_{-\infty}^{\infty} \int_{-\infty}^{\infty} e^{-i(u\xi + vn)} \cdot b(\xi, n, \zeta) d\xi dn = h(u,v,\zeta),$$

$$\int_{z_1}^{\infty} h(u,v,\zeta) \, e^{-\zeta w} d\zeta = F_2(u,v,w) \, .$$

Repeating the reasoning of the first section, we again arrive at the conclusion that the inverse problem for equation (1) in a half-space has at most one solution for which the following representation is valid:

(5a)
$$b(\xi, n, \zeta) = \frac{1}{(2\pi)^2} \int_{-\infty}^{\infty} \int_{-\infty}^{\infty} h(u,v,\zeta) \, e^{i(u\xi + vn)} dudv \, ,$$

$$h(u,v,t+z_1) = \lim_{n \to \infty} \frac{(-1)^n (\frac{n}{t})^{n+1} \frac{\partial^n}{\partial w^n} F_3(u,v,\frac{n}{t})}{n!} \, ,$$

$$F_3(u,v,w) = e^{z_1 w} F_2(u,v,w) \, .$$

Thus the only difference between the inverse problem for a half-space and the corresponding problem for the half-plane is that we have made use of excess information.

Namely, to determine a function of three variables $b(\xi, n, \zeta)$, we employ a function of four variables $f(x_1, x_2, y_1, y_2)$.

The case where the given function is $f(x_1, x_2, y_1)$ has to be investigated further.

BIBLIOGRAPHY

Starred items are in Russian

[1] AGRANOVIČ, Z.S. and MARČENKO, V.A., The Inverse Problem of Scatter-
            ring Theory, Gordon and Breach Science Publishers, New
            York 1963.

*[2] ALEKSEEV, A.S., Some inverse problems in wave propagation theory,
            Izv. Akad. Nauk SSSR Ser. Geofiz., 11, 1962,
            pp. 1514-1531.

*[3] BELONOSOVA, A.V. and ALEKSEEV, A.S., On a version of the inverse
            kinematic problem for a two-dimensional continuous in-
            homogeneous medium. In the collection, Methods and Algo-
            rithms for Interpreting Geophysical Data, "Nauka",
            Moscow 1967.

*[4] BEREZANSKII, Ju.M., A uniqueness theorem in the inverse spectral
            problem for Schrödinger's equation, Trudy Moskov. Mat.
            Obšč., 7, 1968.

*[5] BONČKOVSKII, V.F., The Internal Structure of the Earth, Izd. Akad.
            Nauk SSSR, Moscow 1953.

[6] COURANT-HILBERT, Methods of Mathematical Physics, Vol.2, Interscience
            New York 1962.

*[7] DITKIN, V.A. and PRUDNIKOV, A.P., Integral Transformations and
            Operational Calculus, Fizmatgiz, Moscow 1961.

[8] FADDEEV, L.D., The inverse problem of quantum scattering theory,
            J. Math. Phys., 4, 1, 1963, pp. 72-104.

*[9] FADDEEV, L.D., Increasing solutions of Schrödinger's equation,
            Dokl. Akad. Nauk SSSR, 165, 3, 1965.

*[10]FADDEEV, L.D., Factorization of the S-matrix for the multidimensional
            Schrödinger operator, Dokl.Akad. Nauk SSSR, 167, 1, 1966.

[11]GEL'FAND, I.M. and ŠILOV, G.E., Generalized Functions, Vol.1: Pro-
            perties and Operations, Academic Press, New York 1964.

[12] GEL'FAND, I.M., GRAEV, M.I. and VILENKIN, N.Ja., <u>Generalized</u>
<u>Functions</u>, Vol.5: <u>Integral Geometry and Representation</u>
<u>Theory</u>, Academic Press, New York 1965.

*[13] GEL'FAND, I.M. and LEVITAN, B.M., <u>On the determination of a differ-</u>
<u>ential equation from its spectral function</u>, Izv. Akad.
Nauk SSSR Ser. Mat., 15, 1951, pp. 309-360.

[14] GUTENBERG, B., Physics of the Earth's Interior, Academic Press,
New York 1959.

[15] HERGLOTZ, G., <u>Über die Elastizität der Erde bei Berücksichtigung</u>
<u>ihrer variablen Dichte</u>, Z. für Math. Phys., 52, 3, 1905,
pp. 275-299.

*[16] IVANOV, V.K., <u>On ill-posed problems</u>, Mat. Sb., 61, 103, 1963.

*[17] IVANOV, V.K., <u>Integral equations of first kind and approximate</u>
<u>solution of the inverse potential problem</u>, Dokl. Akad.
Nauk SSSR, 142, 5, 1962.

[18] JOHN, F., <u>Plane Waves and Spherical Means Applied to Partial</u>
<u>Differential Equations</u>, Interscience, New York 1955.

*[19] KREIN, M.G., <u>On the transition function for a one-dimensional</u>
<u>boundary value problem of second order</u>, Dokl. Akad. Nauk
SSSR, 88, 1953, pp.405-408.

[20] LAVRENT'EV, M.M., <u>Some Improperly Posed Problems of Mathematical</u>
<u>Physics</u>, Springer-Verlag, New York 1967.

*[21] LAVRENT'EV, M.M., <u>On a class of inverse problems for differential</u>
<u>equations</u>, Dokl. Akad. Nauk SSSR, 160, 1, 1965, pp.32-35.

*[22] LAVRENT'EV, M.M. and ROMANOV, V.G., <u>On three linearized inverse</u>
<u>problems for hyperbolic equations</u>, Dokl. Akad. Nauk SSSR,
171, 6, 1966.

*[23] MARČENKO, V.A., <u>Some questions in the theory of second order linear</u>
<u>differential operators for one independent variable</u>,
Trudy Moskov. Mat. Obšč., 1, 1952.

*[24] MJUNTC, G., <u>Integral Equations</u>, Part I, GTTL, Moscow-Leningrad 1934.

*[25] NOVIKOV, P.S., On uniqueness for the inverse problem of potential theory, Dokl. Akad. Nauk SSSR, 19, 1938.

[26] PALEY, R.E.A.C. and WIENER, N., Fourier Transforms in the Complex Domain, Amer. Math. Soc. Coll. Publ., 19, New York 1934.

*[27] PRILENKO, A.I., On uniqueness of a solution to the exterior inverse problem for a Newtonian potential, D.U., 2,1,1966.

*[28] PRILENKO, A.I., On inverse problems in potential theory, D.U. 3,1, 1967.

*[29] RAPPOPORT,I.M.,On a two-dimensional inverse problem in potential theory, Dokl. Akad. Nauk SSSR, 28, 1940.

*[30] ROMANOV, V.G., On the determination of a function from its integrals over ellipsoids of revolution with one fixed focus Dokl. Akad. Nauk SSSR, 173, 4, 1967.

*[31] ROMANOV, V.G., On the determination of a function from its integrals along a family of curves, Sibirsk. Mat. Z., 8, 5, 1967.

*[32] SMIRNOV, V.I., Course of Higher Mathematics, Vols. 2,4, GITTL, Moscow 1953, 1957.

*[33] SRETENSKII, L.N., Theory of Newtonian Potential, Gostehizdat, Moscow-Leningrad 1946.

*[34] TIHONOV, A.N., On the solution of ill-posed problems and the method of regularization, Dokl. Akad. Nauk SSSR, 151, 1963, 3, pp. 501-504.

*[35] TIHONOV, A.N., On stability of inverse problems, Dokl. Akad. Nauk SSSR, 1944, 39, 5, pp. 195-198.

*[36] TIHONOV, A.N., On the effect of radioactive decay on the temperature of the earth's crust, Izv. Akad. Nauk SSSR, Otd. Mat. Estestv. Nauk 1937.

[37] WIECHERT, E. and ZOEPPRITZ, K., Über Erdbebenwellen, Nachr. Königl. Gesellschaft Wiss. Göttingen, 4, 1907, pp.415-549.

Lecture Notes in Mathematics

Bisher erschienen/Already published

Vol. 1: J. Wermer, Seminar über Funktionen-Algebren. IV, 30 Seiten. 1964. DM 3,80 / $ 1.10

Vol. 2: A. Borel, Cohomologie des espaces localement compacts d'après. J. Leray. IV, 93 pages. 1964. DM 9, – / $ 2.60

Vol. 3: J. F. Adams, Stable Homotopy Theory. Third edition. IV, 78 pages. 1969. DM 8, – / $ 2.20

Vol. 4: M. Arkowitz and C. R. Curjel, Groups of Homotopy Classes. 2nd. revised edition. IV, 36 pages. 1967. DM 4,80 / $ 1.40

Vol. 5: J.-P. Serre, Cohomologie Galoisienne Troisième édition. VIII, 214 pages. 1965. DM 18, – / $ 5.00

Vol. 6: H. Hermes, Term Logic with Choise Operator. III, 55 pages. 1970. DM 6. – / $ 1.70

Vol. 7: Ph. Tondeur, Introduction to Lie Groups and Transformation Groups. Second edition. VIII 176 pages. 1969. DM 14, – / $ 3.80

Vol. 8: G. Fichera, Linear Elliptic Differential Systems and Eigenvalue Problems. IV, 176 pages. 1965. DM 13,50 / $ 3.80

Vol. 9: P. L. Ivănescu, Pseudo-Boolean Programming and Applications. IV, 50 pages. 1965. DM 4,80 / $ 1.40

Vol. 10: H. Lüneburg, Die Suzukigruppen und ihre Geometrien. VI, 111 Seiten. 1965. DM 8, – / $ 2.20

Vol. 11: J.-P. Serre, Algèbre Locale. Multiplicités. Rédigé par P. Gabriel. Seconde édition. VIII, 192 pages. 1965. DM 12, – / $ 3.30

Vol. 12: A. Dold, Halbexakte Homotopiefunktoren. II, 157 Seiten. 1966. DM 12, – / $ 3.30

Vol. 13: E. Thomas, Seminar on Fiber Spaces. IV, 45 pages. 1966. DM 4,80 / $ 1.40

Vol. 14: H. Werner, Vorlesung über Approximationstheorie. IV, 184 Seiten und 12 Seiten Anhang. 1966. DM 14, – / $ 3.90

Vol. 15: F. Oort, Commutative Group Schemes. VI, 133 pages. 1966. DM 9,80 / $ 2.70

Vol. 16: J. Pfanzagl and W. Pierlo, Compact Systems of Sets. IV, 48 pages. 1966. DM 5,80 / $ 1.60

Vol. 17: C. Müller, Spherical Harmonics. IV, 46 pages. 1966. DM 5, – / $ 1.40

Vol. 18: H.-B. Brinkmann und D. Puppe, Kategorien und Funktoren. XII, 107 Seiten, 1966. DM 8, – / $ 2.20

Vol. 19: G. Stolzenberg, Volumes, Limits and Extensions of Analytic Varieties. IV, 45 pages. 1966. DM 5,40 / $ 1.50

Vol. 20: R. Hartshorne, Residues and Duality. VIII, 423 pages. 1966. DM 20, – / $ 5.50

Vol. 21: Seminar on Complex Multiplication. By A. Borel, S. Chowla, C. S. Herz, K. Iwasawa, J.-P. Serre. IV, 102 pages. 1966. DM 8, – / $ 2.20

Vol. 22: H. Bauer, Harmonische Räume und ihre Potentialtheorie. IV, 175 Seiten. 1966. DM 14, – / $ 3.90

Vol. 23: P. L. Ivănescu and S. Rudeanu, Pseudo-Boolean Methods for Bivalent Programming. 120 pages. 1966. DM 10, – / $ 2.80

Vol. 24: J. Lambek, Completions of Categories. IV, 69 pages. 1966. DM 6,80 / $ 1.90

Vol. 25: R. Narasimhan, Introduction to the Theory of Analytic Spaces. IV, 143 pages. 1966. DM 10, – / $ 2.80

Vol. 26: P.-A. Meyer, Processus de Markov. IV, 190 pages. 1967. DM 15, – / $ 4.20

Vol. 27: H. P. Künzi und S. T. Tan, Lineare Optimierung großer Systeme. VI, 121 Seiten. 1966. DM 12, – / $ 3.30

Vol. 28: P. E. Conner and E. E. Floyd, The Relation of Cobordism to K-Theories. VIII, 112 pages. 1966. DM 9,80 / $ 2.70

Vol. 29: K. Chandrasekharan, Einführung in die Analytische Zahlentheorie. VI, 199 Seiten. 1966. DM 16,80 / $ 4.70

Vol. 30: A. Frölicher and W. Bucher, Calculus in Vector Spaces without Norm. X, 146 pages. 1966. DM 12, – / $ 3.30

Vol. 31: Symposium on Probability Methods in Analysis. Chairman. D. A. Kappos.IV, 329 pages. 1967. DM 20, – / $ 5.50

Vol. 32: M. André, Méthode Simpliciale en Algèbre Homologique et Algèbre Commutative. IV, 122 pages. 1967. DM 12, – / $ 3.30

Vol. 33: G. I. Targonski, Seminar on Functional Operators and Equations. IV, 110 pages. 1967. DM 10, – / $ 2.80

Vol. 34: G. E. Bredon, Equivariant Cohomology Theories. VI, 64 pages. 1967. DM 6,80 / $ 1.90

Vol. 35: N. P. Bhatia and G. P. Szegö, Dynamical Systems. Stability Theory and Applications. VI, 416 pages. 1967. DM 24, – / $ 6.60

Vol. 36: A. Borel, Topics in the Homology Theory of Fibre Bundles. VI, 95 pages. 1967. DM 9, – / $ 2.50

Vol. 37: R. B. Jensen, Modelle der Mengenlehre. X, 176 Seiten. 1967. DM 14, – / $ 3.90

Vol. 38: R. Berger, R. Kiehl, E. Kunz und H.-J. Nastold, Differentialrechnung in der analytischen Geometrie IV, 134 Seiten. 1967 DM 12, – / $ 3.30

Vol. 39: Séminaire de Probabilités I. II, 189 pages. 1967. DM 14, – / $ 3.90

Vol. 40: J. Tits, Tabellen zu den einfachen Lie Gruppen und ihren Darstellungen. VI, 53 Seiten. 1967. DM 6.80 / $ 1.90

Vol. 41: A. Grothendieck, Local Cohomology. VI, 106 pages. 1967. DM 10, – / $ 2.80

Vol. 42: J. F. Berglund and K. H. Hofmann, Compact Semitopological Semigroups and Weakly Almost Periodic Functions. VI, 160 pages. 1967. DM 12, – / $ 3.30

Vol. 43: D. G. Quillen, Homotopical Algebra. VI, 157 pages. 1967. DM 14, – / $ 3.90

Vol. 44: K. Urbanik, Lectures on Prediction Theory. IV, 50 pages. 1967. DM 5,80 / $ 1.60

Vol. 45: A. Wilansky, Topics in Functional Analysis. VI, 102 pages. 1967. DM 9,60 / $ 2.70

Vol. 46: P. E. Conner, Seminar on Periodic Maps.IV, 116 pages. 1967. DM 10,60 / $ 3.00

Vol. 47: Reports of the Midwest Category Seminar I. IV, 181 pages. 1967. DM 14,80 / $ 4.10

Vol. 48: G. de Rham, S. Maumary et M. A. Kervaire, Torsion et Type Simple d'Homotopie. IV, 101 pages. 1967. DM 9,60 / $ 2.70

Vol. 49: C. Faith, Lectures on Injective Modules and Quotient Rings. XVI, 140 pages. 1967. DM 12,80 / $ 3.60

Vol. 50: L. Zalcman, Analytic Capacity and Rational Approximation. VI, 155 pages. 1968. DM 13.20 / $ 3.70

Vol. 51: Séminaire de Probabilités II. IV, 199 pages. 1968. DM 14, – / $ 3.90

Vol. 52: D. J. Simms, Lie Groups and Quantum Mechanics. IV, 90 pages. 1968. DM 8, – / $ 2.20

Vol. 53: J. Cerf, Sur les difféomorphismes de la sphère de dimension trois (Γ = O). XII, 133 pages. 1968. DM 12, – / $ 3.30

Vol. 54: G. Shimura, Automorphic Functions and Number Theory. VI, 69 pages. 1968. DM 8, – / $ 2.20

Vol. 55: D. Gromoll, W. Klingenberg und W. Meyer, Riemannsche Geometrie im Großen. VI, 287 Seiten. 1968. DM 14, – / $ 5.50

Vol. 56: K. Floret und J. Wloka, Einführung in die Theorie der lokalkonvexen Räume. VIII, 194 Seiten. 1968. DM 16, – / $ 4.40

Vol. 57: F. Hirzebruch and K. H. Mayer, O (n)-Mannigfaltigkeiten, exotische Sphären und Singularitäten. IV, 132 Seiten. 1968. DM 10,80 / $ 3.00

Vol. 58: Kuramochi Boundaries of Riemann Surfaces. IV, 102 pages. 1968. DM 9,60 / $ 2.70

Vol. 59: K. Jänich, Differenzierbare G-Mannigfaltigkeiten. VI, 89 Seiten. 1968. DM 8, – / $ 2.20

Vol. 60: Seminar on Differential Equations and Dynamical Systems. Edited by G. S. Jones. VI, 106 pages. 1968. DM 9,60 / $ 2.70

Vol. 61: Reports of the Midwest Category Seminar II. IV, 91 pages. 1968. DM 9,60 / $ 2.70

Vol. 62: Harish-Chandra, Automorphic Forms on Semisimple Lie Groups X, 138 pages. 1968. DM 14, – / $ 3.90

Vol. 63: F. Albrecht, Topics in Control Theory. IV, 65 pages. 1968. DM 6,80 / $ 1.90

Vol. 64: H. Berens, Interpolationsmethoden zur Behandlung von Approximationsprozessen auf Banachräumen. VI, 90 Seiten. 1968. DM 8, – / $ 2.20

Vol. 65: D. Kölzow, Differentiation von Maßen. XII, 102 Seiten. 1968. DM 8, – / $ 2.20

Vol. 66: D. Ferus, Totale Absolutkrümmung in Differentialgeometrie und -topologie. VI, 85 Seiten. 1968. DM 8, – / $ 2.20

Vol. 67: F. Kamber and P. Tondeur, Flat Manifolds. IV, 53 pages. 1968. DM 5,80 / $ 1.60

Vol. 68: N. Boboc et P. Mustață, Espaces harmoniques associés aux opérateurs différentiels linéaires du second ordre de type elliptique. VI, 95 pages. 1968. DM 8,60 / $ 2.40

Vol. 69: Seminar über Potentialtheorie. Herausgegeben von H. Bauer. VI, 180 Seiten. 1968. DM 14,80 / $ 4.10

Vol. 70: Proceedings of the Summer School in Logic. Edited by M. H. Löb. IV, 331 pages. 1968. DM 20, – / $ 5.50

Vol. 71: Séminaire Pierre Lelong (Analyse), Annee 1967 – 1968. VI, 190 pages. 1968. DM 14, – / $ 3.90

Bitte wenden / Continued